Birds of the Australian Rainforests

Paintings by Robert Edden
Text by Walter Boles

REED

Acknowledgements

Author: For critical readings, discussion, and information, I would like to thank Wayne Longmore, Terry Lindsey, Ian McAllan, Tracey Armstrong, Jim Shields, Richard Schodde and Judith Gillespie. I am also grateful to the National Photographic Index of Australian Birds. I would add particular thanks to Lynne Albertson, who prepared the distribution maps that are included in the text. *W.B.*

Artist: I thank Jim Ralston, Ken McCrimmon, John Crowhurst, Bill and Agnes Felton, and Dave Cowan for their help on my field trips for this book. I also thank the Australian Museum for access to its bird study skins and the Australian Bird Park at Palm Cove, Cairns for exceptional co-operation. I make particular mention of Fred Barrie, a valued comrade, with whom I have tramped the rainforests for many, many years. I thank author Walter Boles and Phillip Mathews, who completed a team with which I found it such a pleasure to work. Lastly, I thank my wife May for her encouragement and help throughout this long project. *R.E.*

First published 1986 by
REED BOOKS PTY LTD
2 Aquatic Drive Frenchs Forest NSW 2086

Text copyright Walter E. Boles
Artwork copyright Robert Edden

National Library of Australia
Cataloguing-in-Publication data
Edden, Robert, 1927–
Birds of the Australian rainforests.
Includes index.
ISBN 0 7301 0155 X.
1. Forest birds – Australia.   2. Forest birds – Australia
– Pictorial works.   I. Boles, Walter.   II. Title.
598.29'152'0994
Produced in Australia
Printed in Singapore

Contents

List of Paintings

(shown in order of appearance: see also index)

RAINFORESTS, MAN AND BIRDS

Walter E. Boles

One of Australia's most controversial environmental issues in recent years has been the survival of its rainforests. Attitudes towards this habitat varies. At one end of the spectrum many of the best or most representative areas of rainforests have been preserved in parks and reserves. At the other end lies its destruction for timber or just space for agricultural practices and the encroachment of human settlement with its attendant activities. Such conflict will continue until a balance is found between short-term needs and an appreciation of what this special biological community has to offer the future. It is both necessary and desirable that the long range view is properly considered.

This problem is certainly not confined to Australia; indeed,if the destruction of the world's rainforest continues at its present incredible rate, rainforests may cease to exist anywhere within 50 years. With increasing demands for space, timber and other forest products, even these rates are accelerating. Most countries in which rainforest still exists have unfortunately yet to develop the sort of positive attitudes towards environmental conservation that exists in Australia. This portends a grim future for rainforests in such parts, coupled as it is with the demands of growing populations. Of the 2,000 million hectares of rainforest now on earth, Australia has a mere 0.2 per cent, or about two million hectares. When European man arrived here there was eight million hectares of rainforest, or 75 per cent more than survives today.

Characteristics of rainforests

The majority of rainforests are distributed through the tropical belt of the world. Tropical 'jungles', glorified in so many adventure films, are best represented outside Australia. This country has lush, tropical areas of typical rainforest but other plant communities which are less typical examples of this habitat also fall into this general category. The following points focus on features characterising rainforest and cover most of the variation of this habitat in Australia.

Tropical rainforests are the richest terrestrial habitat in the world; they support the greatest amount of living material of any vegetation community, and add to it at one of the highest rates of any area on earth. It has been estimated that half of all living plants and animals occur in rainforests, many of which have yet to be discovered.

The crowns of rainforest trees touch or overlap, forming an almost continuous canopy cover, from which comes another name for this habitat, closed forest. The canopy may be tall or stunted. Some high-altitude rainforests have canopies of no more than three metres. The principal tree species of rainforests have broad leaves, many very similar looking though unrelated. A characteristic feature is a 'drip-tip' — an elongated tip to the leaf which helps to run off water from the surface. The rainforest trees may retain their leaves throughout the year (evergreen), or lose them in the dry periods to regain them only with the onset of the rainy season (raingreen). Because of Australia's wet-dry seasonality, raingreen forests do not display some of the characteristics of rainforest until sufficient moisture is present.

Below the unbroken canopy may be one or more additional layers. The further into the tropics, the more structurally complex (the more layered) the forest becomes. Extremely large trees which break through to rise substantially above the upper canopy are called emergents.

One of the intriguing aspects of the rainforests is the myriad of unusual types of plants and specialised structures present. Vines are ubiquitous in many rainforests, including thick woody vines, known as lianes, which curl and spiral their way around trunks, or when their support collapses, hang from branches or twist across the forest floor like ropes. Epiphytes are plants able to grow on the sides of trunks, branches and rocks without contact between their roots and the ground, ordinarily without harming the plants which physically support them. Many flowering plants, ferns and mosses exist in epiphytic form, including the well-known Bird's Nest Fern, Staghorn Fern and numerous orchids. Other species, called stranglers, can harm their hosts. From its site high in the tree, a strangler sends roots down its host's side. Once rooted, it begins to grow rapidly in size and may eventually kill the supporting tree by cutting off its resources. When the dead tree finally rots, the strangler must be sturdy enough to stand by itself; if not, it crashes to the ground. The strangler fig is the best known example.

Many of the larger trees of the forest have specialised root systems which form laterally flattened sheets — plank buttresses — extending from the ground up the base of the tree. Several species of trees are characterised by plank buttresses and it is assumed that these contribute some degree of support.

Ferns are often present in the rainforests in a variety of forms, including the familiar Bracken and Maiden-hair ferns, large woody tree ferns, and as creepers or epiphytes. Rainforest palms vary in their density depending on the forest but they are one of the dominant tree forms in some areas. There are numerous species, including Walking-stick and Bangalow palms, impressive fan palms and the notorious Lawyer Vine or 'wait a while', with its formidable array of small but persistent hooks and thorns.

A major source of confusion in defining rainforests has been the presence of non-rainforest plants in an otherwise typical rainforest habitat. Like other plant communities, rainforests go through successional stages involving changes in species composition before reaching maturity. Plant associations of mature rainforest are unrelated to those in the bordering eucalypt forests, yet some species, such as eucalypts and Brush Box, may be present, particularly as emergents. Their existence in low numbers is now recognised as evidence of the transition and growth of such forests. Their occurrence no longer precludes a habitat from the category of rainforest. If they were absent, the remaining forest structure would meet the criteria of rainforest.

Regeneration to the same level of structural complexity and species diversity as the original fo-rest, before logging, may take hundreds of years. Several stages are involved in returning to this mature state. Different species have varying tolerances of exposure: those most resistant to light grow first. providing shelter for other plants which may require greater protection or which may grow more slowly. Some non-rainforest species (eucalypts, for example) act in this capacity and then remain as atypical emergents in the rainforest. The species of each successive stage must be more tolerant of shade as seedlings, but eventually they will break through and outgrow their predecessors. The characteristic canopy trees come last. These need the cover supplied by pioneer species until of sufficient density and size to gain protection from their own crowns. Even an unlogged rainforest is a mosaic of undisturbed areas and natural breaks where the forest may be at various stages in this sequence. Unfortunately the effects of past disturbances may direct the recovery down a new path, leading to the growth of a forest community different to the original.

Types of rainforest

Rainforests in Australia extend through wider latitudes — from the tropics to the temperate zone — than anywhere else in the world. This range of climatic and environmental conditions supports rainforest communities that are not immediately recognised as such by many.

Classification of Australia's rainforests, combining many biological (leaf size, structural complexity, evergreenness, etc.) and environmental (soil fertility, moisture, temperature, etc.) characters, reveals more than 20 identifiable types, including some restricted and specialised communities. Australian rainforests can be grouped into more general and convenient categories (tropical, subtropical, warm temperate and cool temperate) based on mean annual temperature and rainfall. Temperature is closely correlated with latitude, though an increase in altitude has the same effect as an increase in latitude. Within these categories there is a continuum from wet to dry conditions, which further differentiates forest types. Away from the coast and highlands for instance, where the overall humidity is relatively lower, drier forms of rainforest prevail.

Tropical rainforests are, as the name indicates, restricted to the tropics where the annual temperature averages 23°C or above. Wet closed forests of northern Queensland exemplify what is most people's concept of a rainforest. It is here that most of the defining characteristics of rainforest are most pronounced: large leaves, greatest diversity of special life forms, richest species composition and highest complexity of forest structure.

The dry tropical monsoon forests, or vine thickets, occur at the same latitudes as the humid forests but are at opposite ends of the moisture spectrum. These drier forests show noticeable structural distinctions from the wetter rainforests, however, their floristic compositions are more closely related than they are to those of, say, the eucalypt forest and similar habitats. These dry rainforests often have low stunted trees which form a broken canopy; and while most special life forms are greatly reduced, lianes and vines flourish. The trees in this case are raingreen, because they are dependent on those periods of high humidity which alternate with those of seasonal drought. Remnants of a habitat that once was more continuous, dry rainforests survive in scattered pockets, where protection from fire and an accumulation of moisture are greater than in their surroundings.

In the more southern latitudes, lower temperatures (15°C to 23°C annual mean) and a smaller annual rainfall leads to a decrease of many features in the subtropical rainforests. Though its overall richness is still high, it is reduced; special structures exist in smaller numbers and in less complexity. Like the forests of the tropical zone, both wet and dry forms are represented. Among the latter are communities with emergents of native pine and an unusual habitat association of dry acacia (brigalow) and rainforest elements, like bottle trees.

The diminished plant richness and complexity along the latitudinal gradient continues in temperate rainforest, but there are none of the dry communities found to the north. Its trees are always evergreen. Warm temperate rainforest prevails in areas where the annual mean temperature falls between about 12°C and 15°C. This is generally at the foot of mountains and is similar in many respects to cooler stands of subtropical rainforest. In some localities these two types intermingle or grade into each other. The abundance of ferns in warm temperate rainforest has led to it being also known as fern forest.

The cool temperate rainforest is a community of even more southerly latitudes or high altitude, where the mean annual temperature does not exceed 12°C. In this habitat the leaves are small, and the predominance of moss in the cool, moist conditions give it the name 'moss forest'. A few species of tree, sometimes only one, may form the canopy; in most cool temperate forests, these belong to the genus *Nothofagus,* collectively known as Antarctic beeches.

Distribution of Australian rainforests

Older views of rainforest held that the fundamental requirement for its existence was moisture, hence the emphasis on rain in its name. It is now appreciated that a complex interaction of many factors determines the distribution and content of rainforest. These include mean annual rainfall, seasonality of rainfall, mean annual temperature, soil nutrients, soil drainage, rates of evaporation, elevation, annual days of frost, aspect and slope, susceptibility to fire, and others. As these factors change from site to site, they result in a mosaic of forest types, even within a relatively local area.

The total rainforest in Australia, including all wet, dry, hot and cool varieties, is around two million hectares, extending discontinuously through the eastern mainland from Cape York to southern Victoria, in Tasmania, and in far northern Western Australia and the Northern Territory. The greatest variety of rainforest types in Australia is found in Queensland.

Wet tropical forests are limited to areas of the coast and ranges nearby in northern Queensland, as far south as Ingham. Significant patches occur on Cape York Peninsula, the best known around Iron and McIlwraith Ranges. This is the largest remaining area of lowland rainforest in Australia. Because of its remoteness and its protection in reserves, it may be expected to continue to survive largely intact.

Between Cooktown and Ingham is an array of highland and lowland closed forest areas, the most familiar of which is the Atherton Tableland, west of Cairns. Throughout this book this northeastern rainforest tract is often referred to in general terms as the 'Atherton Tableland district', though important sections of rainforest exist away from the Tablelands; this designation is one of convenience in discussing bird distribution. Much has been cleared, leaving isolated fragments in many places. Large valuable rainforest areas remain but timber, agriculture and development concerns continue to threaten them.

A smaller tract of rainforest is found in the Clarke Range, west of Mackay, and in the lowlands around Proserpine, in the middle of the eastern part of Queensland. Dry open regions to the north which extend to the coast segregate it from the Atherton tract.

Dry rainforest is found in scattered areas through subcoastal districts in the east and on Cape York Peninsula, often lining seasonal watercourses. Notable also is the brigalow/bottle tree scrub in the more southern parts of central Queensland.

South of Mackay, smaller patches of wet, closed forest occur — though irregularly — closer to where the borders of southeastern Queensland and northeastern New South Wales meet. In this border area there is a marked overlap of subtropical, warm temperate, and even cool temperate rainforest. Stands of Antarctic beech, for instance, exist throughout the Lamington Plateau district of the Border Ranges as well as in the New England National Park, the higher ranges around the Hastings Valley, and Barrington Tops. Interesting and specialised drier rainforests of this zone include littoral stands on coastal sand dunes and small patches dominated by Hoop Pine emergents. Between the border and the Hunter Valley is a selection of subtropical and warm temperate forests, though these are too often only limited survivors of once larger tracts. Of the Big Scrub in the northeastern corner of New South Wales, what was formerly 75,000 hectares of forests, is now reduced to ten scattered patches totalling a pitiful 300 hectares. On the escarpments south of Sydney, the Illawarra Scrub has suffered a similar fate. Relict pockets of rainforest exist all the way to the Victorian border.

Fragments occur in east Gippsland but most of Victoria's rainforest is in its southern highlands. These cool temperate rainforests are largely dominated by Antarctic beeches, just as they are in western and northeastern Tasmania.

At the opposite end of the continent pockets of monsoon forest are found near the coasts, along watercourses and in some gullies of the sandstone escarpments of Arnhem Land, the Kimberley region, and on Melville Island. Often stunted, they have high proportions of seasonally deciduous trees and large numbers of woody vines. These can survive only where suitable combinations of moisture and fire shelter are found.

Rainforests and man

Since only a quarter of the then existent rainforest has survived European settlement, its effects have been obviously detrimental. Eucalypts, with their hard wood, were unsuitable for many types of specialised woodworking, such as cabinet making or any marine building. Inevitably, settlers turned to the timbers of the rainforests. Red Cedar was the first commercial species to be harvested. Other tree species were adopted as their particular timber properties were discovered. Among these were Sassafrass, Rosewood, Brush Box, Hoop Pine, Huon Pine, Native Teak, White Beech and Coachwood. Such early timber getting required several men to fell a large rainforest tree and teams of bullocks to log them. Technological improvements soon cut the time and cost, and with it, quickened the decline in the integrity of the rainforest.

The most commonly used logging method for rainforest is 'selective logging'. This provides for the retention of a certain proportion of the rainforest canopy, usually half. Under this scheme, the forest structure is supposedly preserved — recovery being promoted through regeneration, while the demand for rainforest timber is met. With a suitable period of rotation between forests, regeneration should, according to its proponents, provide a sustained yield. Unfortunately, these expectations cannot yet be tested practically because the time required for recovery is so long. A sufficient interval after logging has simply not yet passed.

Entire stands were (and still are) removed to provide room for agriculture, such as grazing, banana plantations, or sugar cane fields. The two largest tracts of rainforest in New South Wales at the time of European arrival — the Big Scrub and Illawarra Scrub — are now virtually eliminated, victims of the agricultural push with assistance from the early timber industry. Some cleared areas subsequently proved unsuitable for cattle or crops and were allowed to fall into disuse. Such marginal land remains largely unused in some localities. Other clearing continues to be carried out to make room for the steady encroachment of residential, recreational and tourist development.

Fire rarely penetrates deeply into the rainforest but it attacks it at the edges, killing off plants which act as a buffer. This is particularly significant along boundaries with fire-prone eucalypt forest. The closed canopy protects the trees from the effects of wind. Breaks in the canopy or reduction of the buffering edges, removes this protection and encourages crown dieback. Some trees, unable to cope with prolonged exposure, die or become increasingly susceptible to disease.

As the vegetation is removed, the protection it affords the soil is lost. No longer held by roots, the soil can be washed away by the heavy rains common

<table>
<tr><td>Fan palm.</td><td>Crows Nest fern.</td><td>Staghorn.</td></tr>
</table>

to rainforest areas. The result is severe erosion and a decrease in overall fertility. Exotic plants, such as Lantana, Blackberry and Wild Tobacco, make inroads into newly opened edges and clearings. They establish themselves very quickly and interrupt any recovery by native species. Camphor Laurel, an ornamental tree introduced in the last century, has taken over large areas of rainforests at the expense of native species. Feral animals such as pigs, water buffalo, goats, and rabbits together with uncontrolled domestic stock, disturb the forest floor by causing it to dry out. They disrupt water supplies, destroy the understorey, prevent the regeneration of seedlings and graze at its edges. Cats, dogs and rats have less effect on the plant environment, but they play a sinister role in a vulnerable habitat as highly efficient predators.

These harmful processes are exacerbated by the fragmentation of the forest into isolated patches. In a word: alienation! Alienation has two major consequences beyond the initial habitat loss. Firstly, it separates the resources of each rainforest patch from those of other patches. The further apart they are, the less likely organisms from one can find and make use of the other, the intervening land acting as a barrier across which many plants and animals cannot or will not pass. This can prevent the search for food, shelter or mates. Secondly, the edges of isolated stands are vulnerable to fire, wind and grazing, all of which slowly kill the vegetation, forcing the rain-

forest to retract towards its centre. Eventually the patch may become too small to support the remaining organisms or — less obviously — become inadequate as a habitat in which they can breed successfully. Species, once eliminated from a fragment of forest, cannot be replaced if sources of colonisation are too distant.

Bordering eucalypt forests serve a vital role as buffers for the more sensitive rainforest vegetation. Rainforest cannot be adequately conserved without also providing the protection supplied by surrounding hardwood forests. Rainforests are a source of more than just timber on a commercial scale. Manufacturing products, such as rubber, edible fruits and nuts, like the macadamia, and important medicinal products, including quinine, have been derived from rainforest plants. If demands for rainforest timber, with its special properties, are to be met in the future, some may be cultivated in plantations, like other commercial trees. Reclamation of cleared areas which have proven only marginally valuable for agriculture offer the greatest opportunity. On a world wide scale, it should be noted, nontimber resources have a financially higher return than wood products.

Since many rainforests grow in areas of high rainfall, this habitat serves a valuable role in the protection of watersheds from erosion, nutrient loss and siltation. The tremendous diversity of life in these forests represents a vast genetic reservoir. And it has a largely untapped potential for the development of new commercial and pharmacological products. Rainforest provides a living laboratory, harbouring many of the most primitive members of Australia's plant and animal groups. It also hides a storehouse of information on the evolution of this continent's flora and fauna, and their relationships to those of the rest of the world.

Putting economic and scientific considerations aside, there are still aesthetic, less easily defined and even spiritual values of rainforest that cannot be measured in monetary terms, themselves sufficient to demand its preservation.

Distribution of Australian rainforest birds

The present distribution of rainforests in this country is at most 12,000 years old. The occurrence of rainforest birds is, of course, related to that of the rainforest habitat, but their particular dependence on these areas varies among species. About 15 per cent of Australia's breeding birds are restricted to rainforest or achieve their greatest abundance there. A number of other species, with more flexible habits, freely inhabit both closed and open vegetation communities.

Some 60 million years ago, temperate rainforest covered much of the Australian continent. About this time, earth entered a period of cooling and ice started to build up at the poles. This global loss of free water plus specifically local changes dried out the Australian environment. Falling sea levels exposed a dry link between Australia and New Guinea. Rainforests and their fauna were able to survive only in those areas that could sustain the necessary climatic requirements (refuges) — the mountain ranges of eastern Australia and central New Guinea, parts of Tasmania, and to a lesser extent, the sandstone escarpments of the north, and the southwestern corner of Western Australia.

As conditions oscillated, rainforests went through periods of advance and retreat, perhaps six to eight times, causing large scale extinctions. Birds that still exist in the rainforest refuges of eastern Australia and highland New Guinea are closest to those ancestral species which survived these drastic environmental changes. They are separated from each other by barriers of unsuitably dry land. Forms of birds that are restricted to a single block are known as endemics.

In northern Cape York Peninsula, the lowland rainforests support an array of birds found nowhere else in Australia. They are, however, widespread in similar habitats in New Guinea. This includes the Palm Cockatoo, Eclectus and Red-cheeked Parrots, Blue-breasted Pitta, Green-backed Honeyeater and Magnificent Riflebird, to name a few. These are outliers of the former Australia-New Guinea connection, left when the Torres Strait was flooded for the last time, some 6,000 to 8,000 years ago.

Highlands of northeastern Queensland (the Atherton Tableland district) have been a significant refuge for a long time. These support more endemic species of birds than any other of the rainforest refuges. In addition to birds such as Bower's Shrike-thrush, Fernwren, Bridled Honeyeater and Golden Bowerbird, many more widespread species have distinctive local isolates here.

A smaller, less important refuge is that round the Clarke Range. The only endemic species there is the recently discovered Eungella Honeyeater. In fact, several species which occur in both the Atherton district and in the southeastern rainforests are conspicuously absent from this mid-eastern Queensland block (Pale Yellow Robin, logrunners, Yellow-throated Scrubwren, Satin Bowerbird, catbirds, etc.).

There are few endemic bird species in the southern rainforests, too. These areas, which extend from southeast Queensland through eastern New South Wales, have a diverse and widespread representation of birdlife. Among those few that are confined to this region are the lyrebirds and Rufous Scrub-bird, and distinctive subspecies of Marbled Frogmouth and Double-eyed Fig Parrot.

These birds that occupy the highland refuges of the east coast, represent the older rainforest fauna. Many are represented in the central mountains of New Guinea. Those of the Australian region with closely related forms in New Guinea are: logrunners, Grey-headed Robins, Eastern Whipbird, Brown Thornbill, White-throated Treecreeper and Bridled Honeyeater.

A comparison of the same bird species in different parts of its range shows that southern populations are often less confined to the rainforest alone and are less rigid in their confinement to particular altitudes. Yet the further north we look, the more we find the bird life confined to higher elevations and the more it is likely to remain in the rainforest throughout the year. The endemic species of the Atherton district, for instance, exhibit definite altitudinal preferences for the highlands.

Some principally lowland species of bird are successful throughout the eastern rainforests. Often migratory (e.g. White-tailed Kingfisher, Black-faced Monarch, Rufous Fantail, Spangled Drongo) their greater abilities of dispersal enable them to bridge the gaps that are barriers to other birds. From Cape York to the Atherton district a few of the lowland birds are not migratory (Graceful and Lesser Lewin's Honeyeaters, Boat-billed Flycatcher) but they have nevertheless managed to cross the intervening dry stretches.

The birds of Tasmania's cool temperate rainforests are related to the east coast species. The total number of species in Tasmania is less than in a similar-sized area of the southeastern mainland, so most have been able to exploit a wider range of habitats; birds of the rainforest occur through most types of timbered country.

Birds of the remnant patches of rainforest in northwest Australia are usually distinctive isolates of species found in the rainforests of the east, or closely related forms. These smaller refuges could only support a limited number of species, most as outliers of the formerly contiguous lowland rainforests. Many lowland birds in the north and northeast, such as the Scrubfowl and Black Butcherbird, frequent mangroves as well as rainfo-

rest. Mangroves form a type of closed habitat too, but without the diversity of species or complex structure of rainforest. Despite the difference, however, it provides an important alternative habitat for many rainforest bird species where rainforests no longer exist.

Movements and breeding

Rainforest birds exhibit several patterns of annual movement. One is a regular well-defined north-south migration, as birds depart their breeding grounds for a winter home, returning for the start of the breeding season. Migrants may cross the Torres Strait between Australia and New Guinea or further west to eastern Indonesia (examples: Torres Strait Pigeon, White-tailed Kingfisher, Metallic Starling, Spangled Drongo), or pass between Tasmania and the mainland (Silvereye) across the Bass Strait. Other migratory shifts take place within the continent. Such movements may be expressed by some — but not all — of the members of a species. Southeastern populations commonly migrate towards Queensland in autumn while those already residing in the north remain sedentary. Migrant birds join or bypass the residents.

Species which make only minor movements if at all from the breeding territories are referred to as sedentary. They may defend a territory throughout the year. Nomadism, random movements to find suitable conditions for feeding and breeding, is marked only in the fruit-eating pigeons. As this is the sole constituent of their diet, and the fruiting schedule of rainforest trees is often irregular, these birds are obliged to wander in pursuit of their food. On a more confined scale, local nomadism by some species can take place after breeding. This involves a dispersion from the rainforest into more open habitats in surrounding country or to lower elevations. Immatures often form wandering flocks in autumn. At times of rich nectar sources in neighbouring eucalypt habitats, honeyeaters, parrots and other nectarivorous species leave the rainforest to gather there. Many species make similar opportunistic use of other temporary food resources when they become available.

Though most bird species begin to breed in early spring, there are notable exceptions, such as lyrebirds and logrunners, which begin in midwinter. Year-round breeding is possible in a number of other species but there is a regular peak of reproductive activity in spring and summer. Initiation of breeding is tied to the resources necessary for the

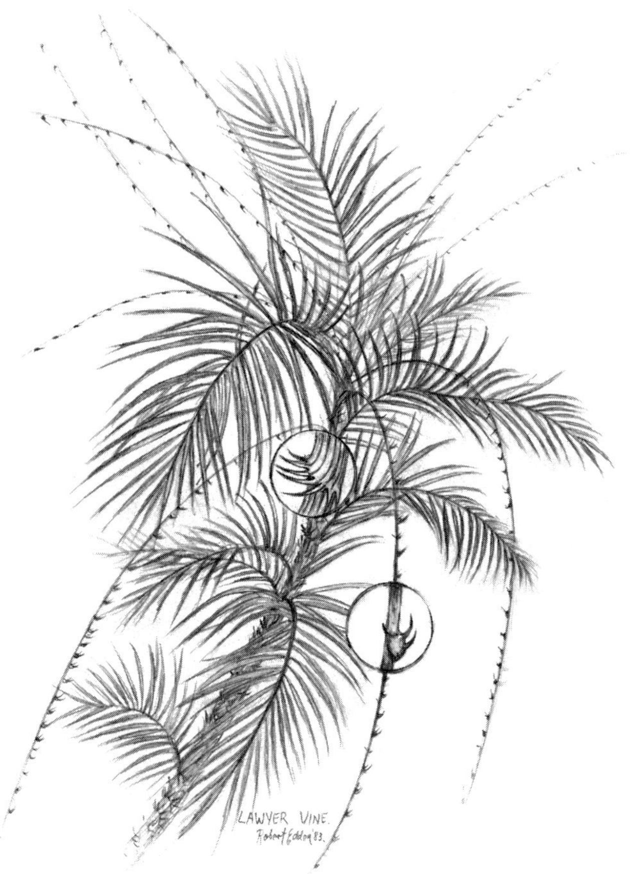

Lawyer vine.

growing offspring. In frugivorous species, for example, breeding commonly coincides with the ripening of fruit. Where there is a definite division of wet and dry seasons, eggs are often laid at the commencement of the rainy period so that an abundance of prey, as a result of the improved conditions, will be available when the chicks hatch.

The incubation period for the eggs depends on the size of the bird and the degree of independence of the chicks once they hatch. It may be a short incubation of around two weeks, or a protracted one, lasting several months. Species that need to nest only briefly because of the self-reliance of their young may produce more than one brood in a season. This is also possible if reproductive behaviour commences early in the year, allowing prolonged nesting for as long as suitable conditions prevail.

Conservation of rainforest birds

The greatest threat facing Australia's rainforest birds (and those of other communities) is the loss of their habitat. While less obvious than shooting or directly killing birds, its effects are more insidious and far ranging. It eliminates the resources on which the entire bird community depends for feeding, roosting and reproduction.

Fruit-bearing trees are essential throughout the year for totally frugivorous birds, like many of the

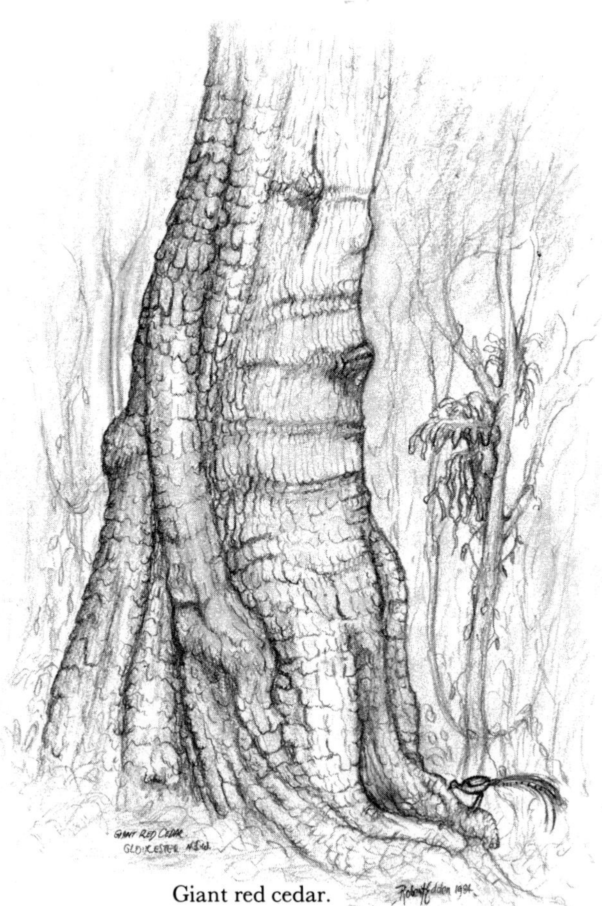

Giant red cedar.

pigeons. The various tree species come to fruit at different times of the year, creating a continual food supply. Disruption of this schedule by removing any one part of this cycle has already proved calamitous for some pigeons. Loss of other food resources may not have immediately obvious effects themselves but the deletion of each reduces the overall potential of the forest to support animal life. The absence of some small but essential item may affect the chances of some birds successfully rearing young.

Migratory species are reliant on stopping-off points in their travels up and down the continent. Again interference with this annual process may lead to larger scale disruptions in their population dynamics.

Perhaps the most significant aspect in which habitat alteration will harmfully affect rainforest birds is through the loss of nesting sites. Parrots, owls and kingfishers, for example, have very definite requirements. They must have hollows of suitable size in the correct locations. Many other rainforest birds need to conceal their nests in the thick undergrowth or canopy. Yet others are simply sensitive to the disturbance of human presence.

Few birds which occur in rainforest — other than the highland endemics of the northeast — are

Giant buttress.

restricted to it. Nonetheless, because at least a portion of their populations obtains a significant part of their resources from this habitat, they may rightly be called rainforest birds. Rainforests and wet eucalypt forests often grade into each other and in some instances it is very difficult to sharply define the border between the two. Eucalypts and rainforest trees mix along their boundaries while the rainforest understorey extends into the open forest. With birds, the delineation is even less definite. They freely move from one to the other, obtaining resources from both. Action to conserve the habitat of rainforest birds must take this into account.

Habitat loss affects the entire bird community, though at the varying rates. The reaction of some species becomes obvious at once but it usually passes unnoticed for some time. Populations can suddenly decrease to critical levels with only the briefest warning. Numbers of many species are still of healthy proportions but further destruction of rainforest will continue to whittle them away.

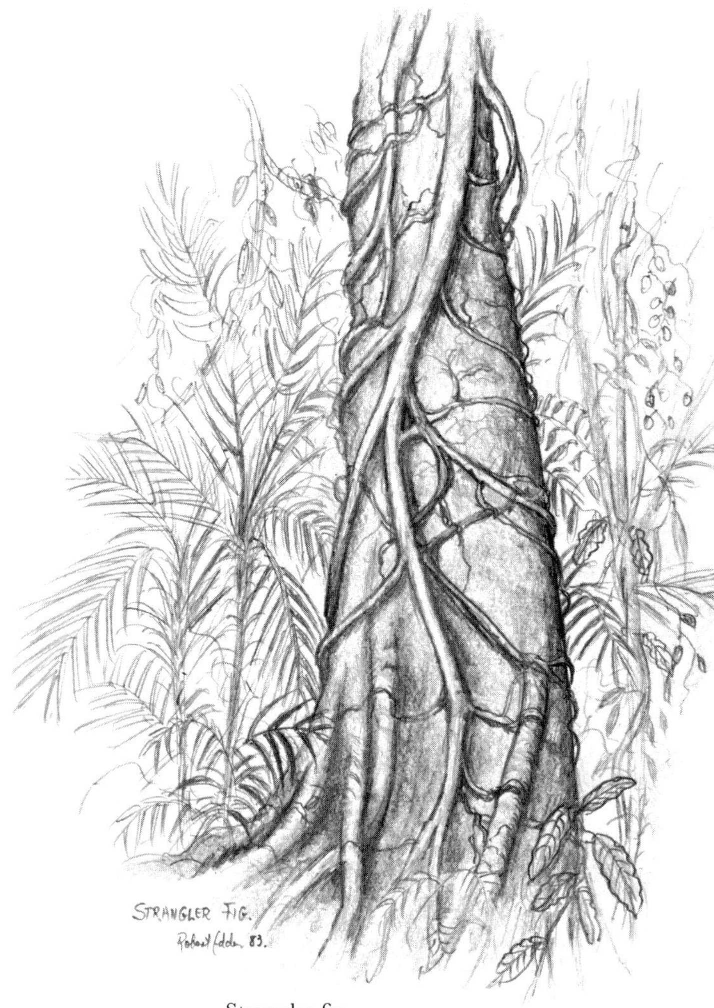

Strangler fig.

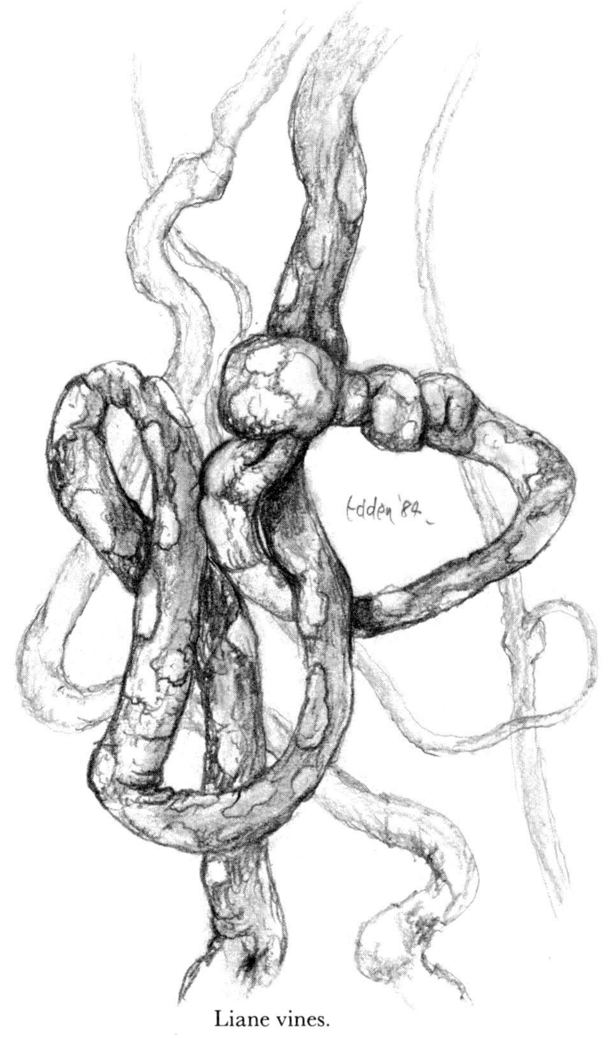

Liane vines.

Shooting and trapping are more noticeable ways in which the birds are harmed and though both are thankfully less prevalent these days, they are not entirely eliminated. Large birds which provide good meat, particularly pigeons, were once killed in considerable numbers. They still are, though on a much smaller scale. Parrots are continually and illegally captured for the pet trade, often with high mortality. In New Guinea several species are now rare round human settlements because of hunting pressure for their meat and feathers.

Our understanding of the more specialised requirements of individual species is not well advanced. Few comprehensive studies have been carried out on the rainforest species — even on the more endangered species. Timber getting may be prohibited in rainforests, which will be a significant step towards their protection, but until our knowledge of each bird's natural history is more complete, their preservation cannot be guaranteed. Perhaps integrated studies of the entire community of rainforest birds is even more necessary. Meantime, more efforts are needed to monitor the response of the birds to man-made changes to their habitat.

Birds of the Australian rainforests...

While a highly representative selection of the rainforest birds of Australia is presented here, it should not be seen as a record of every species that may at some time or other enter this habitat. This is particularly true for birds of Tasmania, many of which occur in all habitats: and of the northwest, where the monsoon and gallery forests play host to a wide range of opportunistic visitors. The various patterns of distribution will be apparent in the range maps. It must be remembered that within these areas, the birds occupy only those parts where suitable habitat exists.

Different names for the same bird often become currency with common usage. Recent efforts to standardise these have not met with universal success. Attempts at changing long-standing names — to make them more 'accurate' or to better reflect current ideas of relationships — sometimes yield replacements even less desirable than the originals. Likewise new proposals on the classification of some species result in changes of scientific nomenclature. Names adopted in this book have been chosen conservatively in most instances but mention of other suggestions is made.

Chapter 1

SOUTHERN CASSOWARY

(Casuariidae; Struthioniformes)

Australia has few land animals large enough to be considered dangerous to humans; one of this small number is the Southern Cassowary *Casuarius casuarius*. It is the second largest bird in Australia, exceeded only by its close relative, the Emu *Dromaius novaehollandiae*. At 1.5 metres to 2 metres in height to the top of its head, the cassowary is shorter than the Emu but outweighs it, reaching nearly 60kg. This, coupled with the cassowary's more aggressive nature makes it a potential threat. In New Guinea these birds are frequently kept as semi-domesticated pets around villages. Their unpredictable and bad temper makes them untrustworthy. Several deaths have been recorded when supposedly tame birds have suddenly attacked villagers. The last record of a person killed by a cassowary in Australia was in 1926, at Mossman, Queensland.

Most of the time, however, cassowaries are shy and, if encountered in the wild, will turn and flee. When cornered or while caring for chicks, a bird may hold its ground and defend itself. The major weapon is a formidable spike-like nail on the innermost of the three toes, 120mm long and 30mm wide at the base. Fighting cassowaries leap into the air, kicking forward simultaneously with both feet, though such engagements between cassowaries are usually brief with little actual damage to the combatants.

The Southern Cassowary belongs to a group of large, flightless birds known as ratites, which include the Ostrich *Struthio camelus*, Emu and rheas (family Rheidae). These are distributed through the southern continents, evidence that these land masses were once united millions of years ago. The ancestors of present day ratites ranged across Gondwanaland — the name given to the land mass of these then connected continents. The Southern Cassowary also occurs in mid-altitude zones of New Guinea. Two other species live in New Guinea as well: the Single-wattled Cassowary *C. unappendiculatus* and the

Dwarf or Bennett's Cassowary *C. bennetti*. The latter also inhabits New Britain on the Bismarck Archipelago, though it is thought to have been transported there by native man.

The Southern Cassowary is a bird of striking appearance. Its large size and powerful legs are only two of its distinctive features. The body of the adult is covered with coarse, hair-like black feathers which lack the barbules that hold the feathers of most birds in their characteristic shape. Each feather has two shafts originating from the same spot on the skin. This unusual structure helps resist wear from thorny or abrasive plants. Brightly coloured skin covers the naked neck and two red wattles hang from its throat,

Cassowary chick.

hence its alternative name, Double-wattled Cassowary.

On top of the head is a large bony helmet or casque, 15cm high, which apparently helps protect the head as the cassowary moves through the thick scrub. When running, it thrusts its neck forward, using the helmet to part the vegetation. Despite its size, the cassowary moves easily through the tangled undergrowth of the rainforest, even when travelling at high speed. Streams are no obstacle. This bird readily enters the water to bathe and it is a good swimmer.

In Australia, rainforest edges, streams and clearings are the chosen habitat of the Southern Cassowary. (In New Guinea it has also been recorded inhabiting mid-altitude savannah.) This species is not uncommon in undisturbed rainforest but it is difficult to locate where there have been intrusions. Certain localities are noted for the relative ease with which a cassowary may be found there. This is because these birds are sedentary, apparently maintaining small territories throughout the year.

Cassowaries remain solitary for most of the year, going their own way in search of food. Fallen fruit is by far their most important food; close to 70 species of fruit have been recorded in their diet. Some is secured from low hanging branches and occasionally cassowaries will make forays out of the rainforest in search of cultivated fruit. Small animals, including carcasses, are also eaten. There is even a

Distribution Southern Cassowary.

record of a cassowary stealing chickens from a hen house.

Should two male cassowaries meet outside the breeding season, they fluff their feathers and, stretching their bodies upwards, produce a rumbling noise. One bird will usually admit defeat to the

other, thus obviating the need to resort to any physical contact. An encounter between a male and female is a different sort of interaction. Like other ratites — but unlike most birds — cassowaries have a

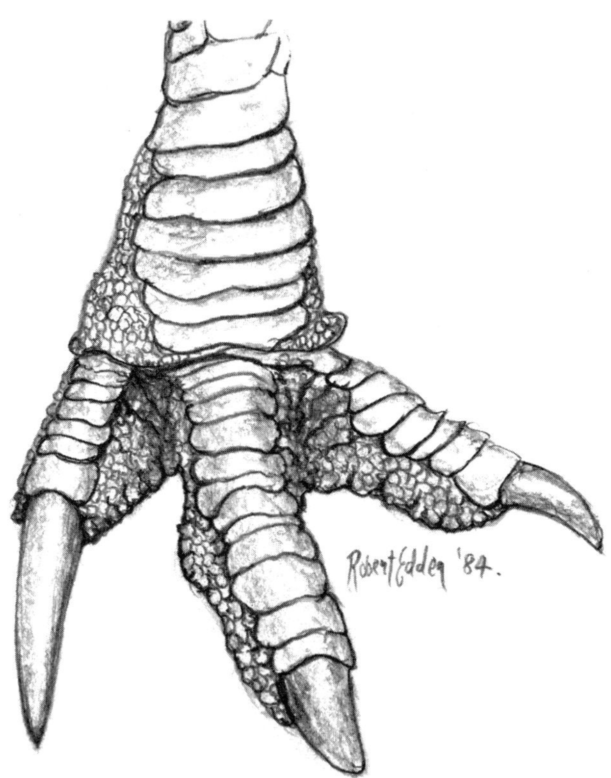

Dagger toe of the Cassowary.

reversal of the roles in the sexes. The female is taller, up to 20kg heavier and more brightly coloured than the male. She is the dominant sex and for most of the year he will turn and flee from her. At the onset of the breeding season (about June), the female grows more tolerant of his presence. Once he is allowed to feed with her, the male will initiate courtship by circling her and emitting a rumbling noise in his throat. If she accepts, copulation takes place. Eventually the female lays three to five large, pale green eggs in a scrape on the ground which serves as the nest. From this point, all tasks of caring for and raising the young are left to the male. The female wanders off and may take another mate.

The abandoned father incubates the eggs for two months. Chicks hatch with striped downy plumage and small wattles. After three months they start to obtain the brown feathers of the immature and by six months have lost their striped plumage completely. As the young grow, they go from brown to black, the crest begins to form and dull colours of the neck start to brighten. They remain with the father for nine months. Adult plumage is not acquired until they are three years old.

Opposite: An adult male Southern Cassowary *Casuarius casuarius* with two immatures.

Chapter 2

BIRDS OF PREY

(Accipitridae; Falconiformes)

Crested Hawk

The attractive Crested Hawk *Aviceda subcristata* is not uncommon within its range but relatively few people see it. Though tame enough to be approached quite closely, its quiet and unobtrusive nature allows it to pass unnoticed most of the time. Despite the boldly barred underparts, large yellow eyes and distinctive black crest, the Crested Hawk may easily be missed as it sits quietly in a tree. In the air, the heavily barred wings and banded tail are readily recognised, but its slow undulating flight over the canopy does not attract the attention received by other more active birds of prey.

During a short period in the breeding season, however, this retiring disposition changes. Crested Hawks engage in acrobatic courtship flights, soaring and tumbling, circling and swooping in spectacular displays. Throughout these performances, they give loud cries. These hawks become generally more vocal throughout the breeding season. Their mellow 'whee-chou' may be regularly heard.

Breeding can begin as early as September and may continue to March, but the usual season is from October to January. The nest is a shallow structure made of sticks and placed on a horizontal limb, usually with supporting side branches. The parents line the inside of the bowl with green leaves and sticks. A nest may be re-used in more than one season but because of its somewhat flimsy structure, it rarely lasts for more than a year. The female lays three to four, and occasionally five, eggs. These are white with a greenish or bluish tone; some have smeary blotches of brown. Both partners incubate the eggs and defend the nest. Incubation lasts about 33 days and the chicks fledge in about the same time again.

Crested Hawks appear to be sedentary. Pairs or small groups of adults and immatures, presumably members of one family, may be seen together through much of the year. Immatures are recognised by brown mottling on the back. Females in the northwestern population of this species resemble the males; those in the east are browner and paler.

Crested Hawks feed on small animals such as insects, lizards, frogs and small mammals. Their talons are slight, not having been evolved for attacking large prey. These hawks also demonstrate their aerobatic skills when feeding. They dive into the foliage of trees to capture prey or dart from branches in pursuit of flying insects. There have even been observations made of birds hanging upside down while searching for food. The same sorts of food are given to the growing nestlings.

This species is one of six in a group known as bazas, cuckoo-falcons or lizard-hawks, which range from Africa through Asia to Australia. It has recently been proposed that the Australian species be given the unfamiliar and somewhat bizarre name of Pacific Baza. Crested Hawk seems preferable. It both better describes the bird and has long been in use. The species also occurs in New Guinea, the Solomon Islands and the Moluccas. In Australia, it is found in two discrete parts of the country, each population having differences in their habitat preferences. Along the east coast, the Crested Hawk lives in rainforest but is not confined to it. It also inhabits thicker eucalypt forests and woodlands and well-timbered rivers. Northwestern birds frequent monsoon forest, mangroves and timber along waterways. Rarely do they venture further than 350km from the coast.

Grey Goshawk

A bird of prey with quite different habits is the Grey Goshawk *Accipiter novaehollandiae*. This bird belongs to a genus of fierce predators. Armed with strong feet and talons, goshawks are powerful, agile hunters. They often lie in wait for passing animals, dashing out from ambush to attack prey in the air or on the

Opposite: Crested Hawk *Aviceda subcristata*.

ground. Reptiles, small mammals and insects are taken but goshawks are most renowned as bird killers. Their assaults are not limited to wild birds; they invade pigeon coops and terrorise caged birds. The Grey Goshawk is often seen soaring over the forest canopy, its flight profile shows broad, rounded wings that are characteristic of this group of hawks. The piping call, 'yuik yuik yuik' may be heard as it flies. Smaller birds of the forest react with cries of alarm before suddenly falling silent until the predator has passed.

Interestingly, there are two distinct colour phases of this species. One is grey on the upperparts with white underparts and breast lightly barred with grey; this is the bird illustrated. The other phase is a beautiful pure white, sometimes called the White Goshawk. In both phases, the cere at the base of the bill and the legs are yellow. That these two different looking birds are the same species is demonstrated by the fact that they freely interbreed, producing young of either colour. The grey phase predominates in thicker vegetation, such as rainfo-

Grey Goshawk *Accipiter novaehollandiae.*

Distribution
Crested Hawk.

Distribution
Grey Goshawk.

rest and dense eucalypt forest, while white birds are more frequent where the habitat is more open. To the north of Australia (eastern Indonesia, New Guinea, the Solomon Islands) this species is different again with a dark grey back and brown-barred underparts.

Grey Goshawks are aggressive when breeding, particularly when defending their nest and young. In the southeast, breeding commences in July and lasts through December; in the northwest the season is later. Both sexes work to construct a large stick platform high in a living tree, lining the inside with green leaves. The female lays two to three eggs, which may be pale greenish-white or have a covering of reddish markings. With little help from her mate she incubates the eggs for five weeks. After they hatch, the female — which is the larger of the sexes — remains at the nest, and both she and her chicks are fed by the male.

The related, but smaller Collared Sparrowhawk *A. cirrhocephalus* ranges throughout Australia in all types of timbered country, including rainforest. Its habits are similar to those of the Grey Goshawk.

Chapter 3

MEGAPODES OR INCUBATOR BIRDS

(Megapodiidae; Galliformes)

Of the various breeding strategies evolved by birds, one of the most remarkable is that of the megapodes (literally 'big feet') or incubator birds. Also called mound-building birds, these incubate their eggs using heat sources other than their own bodies. Some species use volcanic heat, others that of the sun or of decaying vegetation or a combination of sources. There are 12 species of megapodes, three of which are found in Australia. One of these, the famous Malleefowl *Leipoa ocellata,* is a bird of the semi-arid mallee scrub. The other two, the Brush Turkey *Alectura lathami* and Scrubfowl *Megapodius reinwardt,* occupy rainforests.

Brush Turkey

The robust Brush Turkey lives in humid forests along eastern Australia from the tip of Cape York Peninsula to around Sydney and inland to some of the wetter ranges. It also inhabits some drier inland scrubs such as the Pilliga Scrub in northern central New South Wales and the brigalow of southern central Queensland. It is, however, most often observed in rainforest or adjacent wet sclerophyll. Easily recognised, it has a turkey-like appearance, though the broad tail is laterally flattened. Southern birds have bright yellow wattles around the base of the red neck; on Cape York Peninsula, these wattles are light blue.

Like other Australian megapodes, the Brush Turkey constructs a large incubation mound. The male scratches together fallen vegetation, kicking it backwards with his large feet, clearing an area of forest floor up to 20 metres on each side of the mound. As he works over the area repeatedly, the upper layers of the mound accumulate greater amounts of soil than those below. Because they are re-used for more than one year, mounds can become quite large with each annual addition of material. Some may reach four metres in diameter and two

metres in height. Soon after the mound is constructed, the vegetation inside begins to decay. This causes a rapid rise in the internal temperature of the mound. For a while, it is too hot for the eggs to be placed inside but in a few weeks the temperature drops to 33°C to 35°C, the optimum for the eggs.

A mound owner guards his construction aggressively, chasing away other males and allowing females in only to mate and to deposit eggs. The female lays 18 to 24 white eggs, though they are not laid at once. She will visit the mound every two or three days, laying a single egg each time. She makes a hole in the top of the mound, deposits the egg, and covers them before she leaves.

Meanwhile, the male busies itself maintaining a proper incubation temperature in the mound. He gauges the temperature by digging holes in the mound and inserting his head. If it is too hot, he removes some of the outer layer to allow heat to escape. If too cold, additional material is added to trap and retain more heat. Proper conditions last two to three months.

This temperature regulation is the only assistance the parents give their offspring. Many eggs are no doubt lost to predators such as goannas and feral pigs. Those that survive hatch in about seven weeks. Because the laying of the eggs is staggered, so too is the hatching. Chicks must dig their own way to the surface of the mound and from then on they are entirely on their own. To compensate for this forced independence, the chicks are very precocious when hatched. They emerge fully feathered and are able to run and feed themselves immediately; within a few hours they can fly.

Brush Turkeys remain in the same area throughout the year. When not breeding, they spend the day searching the forest leaf litter for food, which includes fruit, seeds and small animals. Their calls are grunts, fowl-like clucking and a loud deep 'kyok'.

Most Brush Turkeys are usually spotted running quickly across roads. They prefer to escape danger on foot but if pressed, fly heavily, usually moving into the trees. One bird was seen to kick debris at an approaching goanna, perhaps as a primitive type of defence. Generally these birds are wary but around picnic grounds or homesteads they can become extremely tame, almost semi-domesticated. Brush Turkeys may join domestic fowl in farmyards or accept handouts from park visitors. At night they roost in trees.

Brush Turkeys once extended south of Sydney but clearing and shooting have led to local extinction in this and other areas. In much of its range, this species is still common. Some westward expansion was noted in conjunction with the spread of the Prickly Pear cactus, a fruit enjoyed by this bird.

Distribution Brush Turkey.

Scrubfowl

Although it is only half the size of the Brush Turkey, the Scrubfowl builds a larger mound. Usually five metres wide and two metres high, mounds several metres larger are found. Even gigantic 12 metre by 5 metre structures have been recorded. As the season progresses, the mounds shrink in size with the decay of the plant material and constant use. Both sexes maintain the mound. Several pairs of Scrubfowl — a more social species than the Brush Turkey — may use the same mound, making it difficult to determine the clutch size of one female or the duration of incubation. Breeding activity begins in August and persists to March. The eggs are laid at the end of a tunnel up to two metres long, dug into the mound. After they are deposited, the eggs are reburied by the parents. The rotting vegetation stains the eggs, which are yellowish-pink when laid, with a brownish wash.

Scrubfowl *Megapodius reinwardt.*

The occurrence of mounds are an indication of the presence of Scrubfowl, or Junglefowl, as they are often called. Another sign is the loud cry, a weird double crow. Other calls include a clucking like that of a fowl. If startled, Scrubfowl either flee rapidly on their powerful orange legs and feet or fly to a high branch and peer at the intruder. Their pointed little crest gives them a comical appearance.

Scrubfowl live in rainforest, gallery forest and occasionally, mangroves. There are populations on the tip of the Northern Territory, along coastal regions of east Queensland and on islands in Torres Strait. This and closely related species range from the Philippines and Indonesia through New Guinea to the Solomon Islands and Vanuatu. The limits of the various species are not easy to determine. They differ in their breeding strategies, some using conventional mounds, others placing their eggs in burrows, rock fissures warmed by the sun or the sands of beaches. The similarity in appearance of these forms further complicates the problem.

Distribution Scrub Fowl.

Opposite: The Brush Turkey *Alectura lathami.*

Robert Lloder 1983.

Chapter 4

BLACK-BREASTED BUTTON-QUAIL

(Turnicidae; Gruiformes)

Despite its outward appearance and its name, the Black-breasted Button-quail *Turnix melanogaster* is not closely related to the true quail (family Phasianidae). It is considered more closely allied to the rails and cranes. A superficial difference between this species and the quail is its three, rather than four toes (the back toe is missing), but more important are significant anatomical differences. Another notable separation between button-quail and true quail is the reversal of sex roles in the former. The female is the larger of the sexes (190mm compared with 165mm) and it is more boldly coloured than the male who lacks any large, solid black patches in his plumage. It is the female who establishes a territory, defending it against others of her sex. Her deep, booming, rapidly repeated 'oo-oom' attracts potential mates to her territory.

While breeding may take place in any month, the colder months are generally avoided. Pairs in captivity have bred from October to March. Both sexes build the nest, a small depression in the leaf litter on the ground, which is often at the base of a small bush. The eggs — usually two to four — are white with fine speckles of brown and larger blotches of darker blackish-brown. Once the eggs are laid, the male assumes all responsibilities of nest maintenance. He incubates the eggs for 16 days. For several days after the chicks emerge from the eggs, they are fed by the male, but they are soon able to find food for themselves. They grow so rapidly that they are capable of breeding themselves in five to six months.

Immatures resemble the adult male except for their darker legs and eyes.

Black-breasted Button-quail have not been well studied in the wild and much of our information about them comes from captive birds for they are not uncommon in aviaries. The species is considered rare in the wild, though this may be more a reflection of its shy, retiring nature than of its actual abundance. Whatever the reason, this bird is not often seen. Most frequently it is encountered in pairs

Distribution Black-breasted Button-quail.

or small coveys in clearings and along edges of drier rainforest and vine scrubs. Apparently this species is sedentary.

The most common indication of the presence of button-quail is small, saucer-shaped depressions in the forest floor. These are formed by the birds as they feed. Scratching with one foot, they pivot on the other, excavating a shallow semi-circle, then shift feet and reverse direction. Seeds and small invertebrates are their main diet.

Opposite: Black-breasted Button-quail *Turnix melanogaster* — female (above); male (below).

Chapter 5

RAILS

(Rallidae; Gruiformes)

Some of the larger rails, such as coots, swamphens and moorhens, are common and well-known inhabitants of parks, dams and streams. In general, however, the smaller species are noted for their skulking, secretive habits. They move stealthily through waterside vegetation, usually most active at dusk or after dark. Their loud distinctive cries are the most acquaintance that many people ever have with these wary birds. A taped replay of calls will attract rails without great difficulty, rewarding the patient observer who is willing to take time and effort. When flushed, rails prefer to make their escape by swiftly moving through the reeds and waterside grasses but if chased, dive underwater or take to the air with a seemingly weak flight, feet dangling as they go. Despite this impression, this group of birds are among the most successful colonisers of many islands, including those throughout the South Pacific. A number of island species have lost the power of flight since the arrival of the original colonising birds. Australia's 16 breeding species can be found in a variety of wetland and streamside habitats. Two are commonly associated with rainforest, one almost exclusively.

Red-necked Rail

The Red-necked Rail *Rallina tricolor* occurs along waterways in dense rainforest from northeastern Queensland to the Moluccas, New Guinea and the Bismarck Archipelago. It is one of several similar forest-inhabiting rails of this region. The abundance of this bird is hard to gauge because of the difficulty in observing it. In some areas the Red-necked Rail may be common but surveys of its presence would probably have to rely on counting calling individuals. If seen, it cannot be confused with any other species within Australia; the red head, neck and breast, and long-legged body are diagnostic. More often than not its presence is announced by nocturnal calling. This rail is a noisy species, producing a range of strange sounds, some described as resembling the grunting of pigs and of 'berries dropping in the water'. Its other common calls are a persistent and monotonous 'clok clok clok' and a descending 'nark nark nark'.

Many rails have heavily blotched eggs, but those of the Red-necked Rail are, oddly, glossy white. For many years this peculiarity, combined with its secretive habits, made the identification of its eggs tentative. It was not until a bird was observed sitting on a clutch that the description of the eggs could be confirmed. Three to six eggs are laid on the forest floor, among the fallen litter at the base of a tree, or sometimes in a shrub up to half a metre off the ground. Incubating rails lose much of their shyness and do not like leaving the nest, remaining there until closely approached. The chicks, like those of all Australian rails, are covered with black down upon hatching. They are able to run shortly afterwards and feed themselves within a few days. Small freshwater invertebrates found along the stream comprise the principal part of the Red-necked Rail's diet. Foraging birds use their bill to turn over rocks and sticks and grab the exposed prey.

Distribution
Red-necked Rail.

Opposite: Red-necked Rail *Rallina tricolor.*

Robert Madden 1988.

Bush-hen

Less confined to the rainforest is the Bush-hen *Amaurornis olivacea*. It dwells along the edges of rainforest, moving out to other areas of thick, rank vegetation such as swamps, creekside thickets, and even dense gardens. It seldom strays far from water. Despite its shyness the Bush-hen may feed along cleared areas adjacent to thick scrub, such as golf courses, or be seen dashing across a road on its strong legs. Road casualties or cat kills often provide the only views of this elusive bird. So quietly can it crouch in the underbrush that a person passing within a metre is unlikely to detect it.

The calls of the Bush-hen are weird and varied. A pair may call as a duet, giving wild shrieking sounds alternatively compared with that of a cat in pain or donkey-like in character. It may be set off by loud, unexpected sounds. Other calls include clicking notes and grunts. Sexes can be distinguished by differences in the pitch of the voice.

Bush-hens breed from October to April. Early in the season, both parents are very vocal and expend considerable energy guarding their territory from intruders. They trample down a patch of grass in the midst of heavy vegetation with their feet. On this platform, the female places four to seven cream-coloured eggs blotched with brown and lavender. Additional plant material may be placed on the nest and surrounding vegetation pulled over

Distribution Bush-hen.

the top. With the laying of the eggs the vocal level dies down markedly. But it returns to its earlier intensity after they hatch. Like those of the Rednecked Rail, the chicks are black. Soon after they hatch they leave the nest, though for the next week they return to roost for the night. A nest-like structure may be constructed by the adults as a roosting area.

Bush-hen *Amaurornis olivacea*.

The general appearance of the Bush-hen might be mistaken for the more familiar Dusky Moorhen *Gallinula tenebrosa*, a larger, greyer bird. The buff-coloured belly and undertail coverts, however, clearly differentiates the Bush-hen from other species, features that have led to it being called the Rufous-tailed Moorhen in other parts of its range, which includes the Philippines to the Moluccas and east to the Solomon Islands.

The first record of the Bush-hen in Australia appears to be a clutch of eggs collected from New South Wales in 1864; though since then it had only been seen in northeastern Queensland. Recently, however, this species has been recorded with increasing frequency from the area around Brisbane and far northeastern corner of New South Wales. Whether these 'reappearances' represent recently established populations or birds that had always been present but managed to remain undetected is not known. Neither is it known if there are annual movements by the Bush-hen within its Australian range. Its presence in the Northern Territory has only been confirmed in the past few years.

Chapter 6

PIGEONS

(Columbidae; Columbiformes)

Of Australia's 22 species of native pigeons and doves, ten occur regularly in rainforests. The rest prefer more open areas where some appear to fill the ecological role of partridges, which are absent from this continent. Open country forms are terrestrial seed eaters. Those of the rainforest include three semi-terrestrial species which eat both seeds and fruit and seven which are fruit specialists. The frugivorous pigeons are mostly, if not entirely, arboreal. They undergo marked population movements, even between widely separated localities, as they follow the appearance of ripening fruit. This dietary dependence on fruiting trees at all times of the year makes them extremely vulnerable. Removal of the trees through logging, grazing, and general agriculture has already reduced the numbers of some species. Shooting of these edible birds has also taken its toll.

Pigeons have been the subjects of numerous behavioural studies because of their fascinating courtship displays. A male performs a special display flight to attract a mate. He then courts her with a ritual of bowing, cooing and other ceremonial activities. There are many other behaviours and postures, the functions of which are not all understood. Some are involved with threats, defense and comfort movements, as well as breeding. Calls of the species are good recognition clues. Each has a variety of vocalisations, including 'coo's' which vary depending on their meaning and the context in which they are given.

All pigeons lay white eggs. Fruit-doves and other arboreal species usually lay only one but the other rainforest forms may lay two. Nests of some pigeons are almost ridiculously frail, a mere handful of scattered sticks laid haphazardly across a fork in a branch.

Red-crowned Pigeon

Many of the rainforest pigeons are surprisingly brightly coloured, quite unlike the introduced Feral Pigeon *Columba livia* which most people think of as a 'typical' example of this family. One of the most highly coloured is the Red-crowned Pigeon *Ptilinopus regina*. With the following three species it belongs to a genus sometimes collectively called fruit-doves. This group occurs through the South Pacific, west into the eastern islands of Indonesia, reaching its greatest diversity in New Guinea with 14 species. The Red-crowned Pigeon has two disjunct populations in Australia, and it is also found in Timor and surrounding islands.

Male and female Red-crowned Pigeons are similar in apppearance, the female being slightly smaller. Immatures lack much of this colour, including the distinctive rose crown. As brightly patterned as they appear, they are not, as one might think, easy to spot. These pigeons are shy and 'freeze' when an intruder approaches, their green backs blending with the foliage and the coloured crown usually turned away. There are several ways in which these birds can be located. Brief views may be obtained during their quick, direct flight between trees. The sound of falling fruit can signify that they are on a nearby feeding tree. Fruit comprises the entire diet of Red-crowned Pigeons. They are active, even acrobatic feeders and individuals sometimes hang upside down to reach berries or other small fruit. While feeding, these birds are usually silent; at other times, their characteristic calls draw attention to their presence. This may be a soft three-note 'boo-uk-boo' or a series of 20 or more loud 'coo's' which decrease in duration and pitch while accelerating in speed.

Red-crowned Pigeons may be seen entering woodlands and mangroves and may even visit isolated fruit-bearing trees away from the cover of dense forest. This seems true more, however, in the northern part of their range because in the south, Red-crowned Pigeons are more restricted to rainforest. They are locally nomadic throughout their range, moving in pursuit of food. There may be large scale migrations but these are poorly understood. In some areas pigeons may be found throughout the year, in others, numbers vary seasonally in a manner suggestive of a more regular migration.

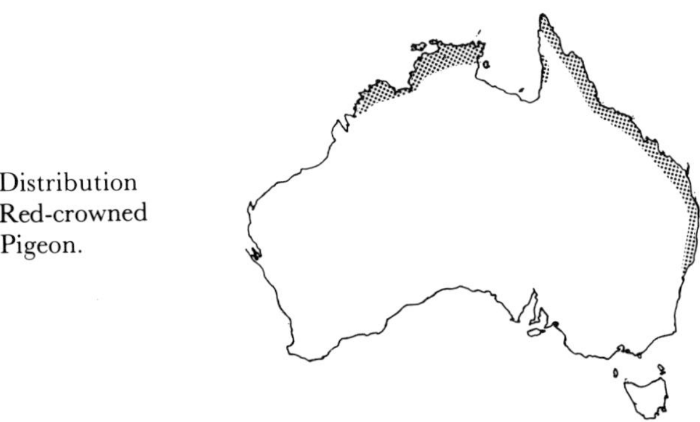

Distribution Red-crowned Pigeon.

Like most species, the Red-crowned Pigeon has a series of complex courtship displays. One, present with variations in most of these species, is the bowing display. The male stands erect, with his bill pressed down against his inflated chest. Keeping his tail hanging motionless, he bends slowly forward until nearly horizontal, then returns to his original position. In another ceremony, called 'peering', the male faces the female and moves his head side to side, while his gaze remains fixed on the subject. While doing this he slowly advances down the branch towards her.

This species has one of the frailest nests of any of the pigeons, little more than a simple platform of a few sticks.

Purple-crowned Pigeon

The Purple-crowned Pigeon *P. superbus,* a bird of similar size, shares much of the same range as the Red-crowned Pigeon. In general habits, these species are close but each has subtle differences that allow them to co-exist. The extra-limital range of the Purple-crowned Pigeon is more extensive (New Guinea, Bismarck Archipelago, eastern Indonesia) but within Australia it is more limited. Occasionally vagrants reach Sydney and, rarely, even Tasmania, but breeding occurs south only to about Proserpine, Queensland. It is a regular visitor to Brisbane. Through its range, this pigeon prefers lowland rainforest or other forest with a thick, rainforest undergrowth.

The movements of this species are not understood. Locally nomadic in the quest for food, some

populations are sedentary while others are possibly migratory. Outside its normal range, it may appear in a variety of habitats. Often these pigeons are found dead, having struck a window of a house; apparently they have difficulty during their swift flight distinguishing a reflected image from open space.

Like its relative, the Purple-crowned Pigeon is shy and difficult to see when perched among the leaves of a tree as it feeds. Over 50 species of fruiting plants have been recorded in its diet. In contrast with the previous species, the Purple-crowned Pigeon shows strong sexual dimorphism, the female lacking many of the showiest colours of the male. Immatures resemble the female. An alternative name, Superb Fruit-dove, comes from the brightly plumaged male.

These Red-crowned and Purple-crowned Pigeons can be separated by calls when the birds themselves cannot be seen. The latter has a loud,

Distribution Purple-crowned Pigeon.

double note 'coo-coo' and a low 'oom', given either singly or in a steady, non-accelerating series.

It is in the breeding biology that these two fruit-doves show the greatest divergence. Breeding of the Purple-crowned Pigeon may take place throughout the year, but in north Queensland is most frequent from June to February; elsewhere November to December is the usual period. The general pattern of the courtship display of this bird is like that of the Red-crowned Pigeon, except for the apparent lack of 'peering' behaviour.

Incubation requires 14 days, four days less than that of the Red-crowned Pigeon. Both parents incubate with the male sitting during the day. Because his bright colours might attract a potential predator, he has an unusual method of concealing these sections of his body. If approached, he raises his tail at

Opposite: Red-crowned Pigeon *Ptilinopus regina.*

Robert Edden 1983.

an intruder. The mottled green and white undertail coverts blend with foliage and hide his more gaudily coloured head and mantle. As the intruder moves, so does the pigeon, always maintaining a position where he can present his undertail to his enemy.

Wompoo Pigeon

Much of the breeding behaviour of the larger Wompoo Pigeon *P. magnificus* has not been reported. Birds are sexually active year round though breeding is concentrated between June and January, with a peak from August to October. The display, nest and egg are similar to those of the previous species though its incubation time is not known. To many observers, this species is the most familiar of the three. Its strange and well known call starts with a short 'cluck' and finishes with a bubbling 'wompoo', hence its name. At a distance, several calling birds sound like human voices carrying on a muffled conversation in the rainforest. A nickname for the Wompoo Pigeon, also derived from its call, is Bubbly Mary. Magnificent Fruit-dove, another alternative, stresses the striking combination of colours.

Wompoo Pigeons occur through lowland New Guinea and down the east coast to about the Hunter Valley district of New South Wales. Formerly they extended to Illawarra, south of Sydney, but shooting and clearing of rainforest has retracted their range. Nowhere in New South Wales is this species common. From south to north it shows a marked decrease in size. Southern birds are over 550mm in length, those of north Queensland, 350mm. Nonetheless, even the northern Wompoo Pigeon is noticeably larger than its two relatives with which it shares much of its Australian distribution. Unlike

Distribution Wompoo Pigeon.

them, this species is sedentary, other than the odd local movements. It, too, eats fruit, the variety varying throughout the year, and from year to year,

depending on the irregular schedule of fruiting within the rainforest.

Banded Pigeon

Along the western edge of the Arnhem Land escarpment in the Northern Territory lives the fourth Australian member of this genus, the Banded Pigeon *P. alligator*. With its black and white plumage, it does not seem obviously related to the other brightly coloured fruit-doves. Closely related birds in nearby Indonesia, with which the Australian bird has sometimes been considered conspecific, bridge this gap. Where the underparts of the Banded Pigeon are grey, the others are olive-green.

This large bird is the scarcest of Australia's pigeons. Perhaps more widespread when rainforest was more abundant in northern Australia, it is now restricted to the stands of gully rainforest along the

Distribution Banded Pigeon.

escarpment. It ranges from Oenpelli, south to around the South Alligator River, but the eastern limit of the distribution has not been determined. Within this area the Banded Pigeon is not uncom-

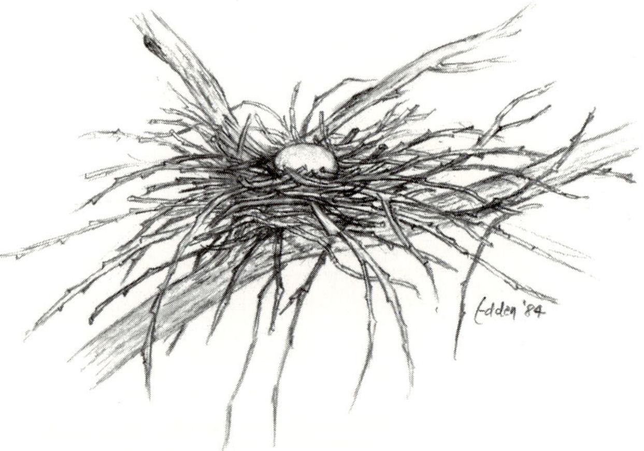

Purple-crowned Pigeon nest.

Opposite: Purple-crowned Pigeon *Ptilinopus superbus* — male (above); female (below).

Robert Edden 1983

mon but is locally distributed. The nest was not discovered until 1971 and other aspects of its natural history are not well known.

Like the other fruit-doves, this species is frugivorous but it is not as agile a feeder. Breeding takes place from April to November, before the onset of the wet season. Display patterns which have been described resemble those of the other Australian fruit-doves. The major call is a loud, low 'coo'.

Although much of this species' range is now protected in Kakadu National Park, its small population and specialised habitat make it vulnerable to environmental disturbances such as mining.

Banded Pigeon *Ptilinopus alligator*

Torres Strait Pigeon

Another fruit-eating pigeon is the black and white Torres Strait Pigeon *Ducula spilorrhoa*. Alternatively called Nutmeg Pigeon or Torresian Imperial Pigeon, it is the only Australian member of a genus which reaches its greatest diversity in New Guinea and surrounding islands. This species has occasionally been recorded from as far south as northern New South Wales, but its normal distribution is lowland New Guinea, the Bismarck Archipelago, and coastal regions of northern and northeastern Australia.

The Torres Strait Pigeon is noted for its conspicuous movements. Australian populations are migratory, moving to New Guinea and back each year. Queensland birds start their northwards journey in January, peaking in February and March. By April, most have departed except for a few which remain throughout the year. The destination of these migrating birds is not fully understood. Individuals banded in Queensland have been recovered in widely separated parts of Irian Jaya and Papua New Guinea; the exact routes followed for this dispersal have yet to be plotted. The destinations of the populations in Western Australia and Northern Territory remain unknown. Migrants return to Queensland in late July, reaching Cape York and moving down both coasts the Peninsula. Numerous flocks of 50 to 100 birds — sometimes up to 500 — can be seen making their way across the Torres Strait close to the water's surface.

Upon returning to Australia, Torres Strait Pigeons begin breeding almost immediately. The majority of these nest on offshore islands, travelling to the mainland each day to feed. Breeding colonies on these islands are often quite large. Many are situated in mangroves but nests have been found on rock ledges, in abandoned nests of other species and even on the ground. Torres Strait Pigeons construct a more substantial nest than do many other pigeons, often consisting of mangrove shoots.

The male has a prominent display flight, undertaken early in the breeding season. He rises steeply in the air, reaches a peak and glides down to his original level. Several of the other displays are like those of the fruit-doves but more exaggerated. During the bowing display, the male inflates his neck and breast. He arches his neck, presses the bill against his breast and rapidly bows up and down.

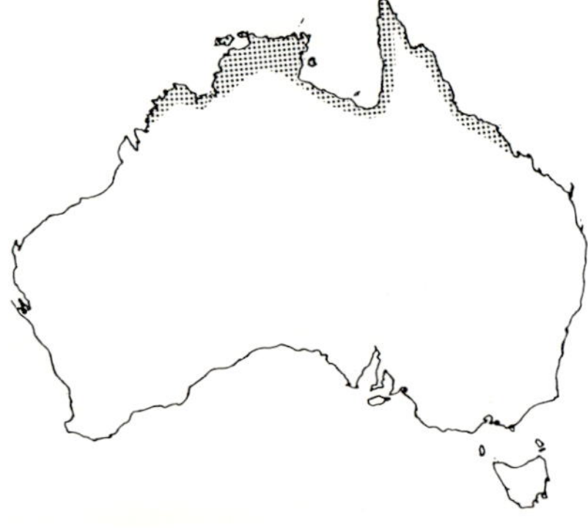

Distribution Torres Strait Pigeon.

Unlike the Red-crowned Pigeon and its relatives, the Torres Strait Pigeon is conspicuous when

Opposite: Wompoo Pigeon *Ptilinopus magnificus*.

moving to and from feeding areas. Breeding colonies are noisy places with birds arriving and departing, and males calling and displaying to their rivals. These pigeons feed in much the same manner as the fruit-doves at which time they are unobtrusive and may go unnoticed. Distribution of colonies of this

Torres Strait Pigeon *Ducula spilorrhoa.*

species is regulated by the presence of rainforest on the adjacent mainland. Torres Strait Pigeons are not often found in the middle of extensive stands of rainforest, preferring instead to forage along the edges. Fruit is their only food. In its search for fruit, this species freely visits eucalypt and paperbark forest, mangroves and riverine scrub.

The calls of the Torres Strait Pigeon include a loud, low 'coo-woo' and a 'coo-hoo-hoo' given during the bowing display.

Because of its colonial habits, this species is particularly sensitive to human disturbance. Reductions in numbers are already marked. Early reports estimated up to 100,000 birds forming some colonies; the largest now numbers around 25,000. Destruction of the feeding habitat, notably the clearing of rainforest for sugar cane, led to a rapid decline in the southern part of this species' range. It has been further affected by shooting of birds at the offshore colonies. Removal of adults at this time also results in the death of the young. Though many of the islands are now protected as reserves, disturbances persist. More importantly, alteration of

rainforest will continue to seriously affect the Torres Strait Pigeon regardless of how secure the breeding islands are.

White-headed Pigeon

Another pigeon which suffered as a result of man's activities was the White-headed Pigeon *Columba leucomela*. In the late 19th century, the clearing of rainforest for agriculture and other purposes, resulted in severe losses of this species. Once rare, it has made a gratifying comeback, primarily because of its ability to utilise new plant species introduced by man. Many cleared areas were planted with the exotic Camphor Laurel. This vigorous tree spread quickly and was soon adopted by the White-headed Pigeon (and other species) as an acceptable alternative source of fruit. As many cleared areas proved unsuccessful for crops and cattle, they were allowed

White-headed Pigeon *Columba leucomela.*

to regenerate with rainforest trees. Gradually there was a recovery in the availability of this pigeon's native food sources. In some localities, maize stubble has served as a suitable supplement to the diet of the White-headed Pigeon. These dietary shifts have released this species from its absolute dependence on the rainforest. It breeds in this habitat but it often leaves the denser scrub to feed in more open situations. White-headed Pigeons may occasionally feed, or even nest, in introduced street trees.

Though this species lives only in Australia, closely related species can be found in the southern islands of Japan, the Philippines, New Guinea and Micronesia. Its nearest relative in Australia is the common introduced Feral Pigeon, a ubiquitous resident of so many city parks. The range of the White-headed Pigeon extends along the coastal regions of eastern Australia and some adjacent highlands, from Cooktown south to the Hunter River district. South of this, records are more scattered. The bird is an irregular visitor to Sydney; it has even extended to Eden on the south coast, and breeding has been recorded at Nowra. In the northern part of its distribution, this pigeon occurs most often at higher elevations than it does elsewhere.

The breeding biology of the White-headed Pigeon is somewhat of a mystery. The few records that exist suggest that the breeding season is extended and with two broods sometimes raised in the same nest in a season. Nesting has been noted in almost all months of the year but seems to be most intensive in October-December.

Other evidence of breeding is the conspicuous courtship flight of the male which he initiates by rising from the canopy and flying in a circular path. At regular intervals he then turns sharply upwards before swooping back again to his original altitude. At the end of the path he returns to the upper layers of the forest. The male White-headed Pigeon also has a bowing display performed in the presence of a female. He stands upright, facing her with his plumage fluffed out and throat inflated. The body is stiffly inclined forward five or six times in succession. At the bottom of each bow, he utters a soft 'coo'.

Distribution White-headed Pigeon.

Calls of the White-headed Pigeon — described as low and mournful — include a drawnout 'oom-coo' and a single 'oom'. These notes are ventrilocal, that is, they appear to originate at locations other than from where the bird is producing them. Finding a calling bird is difficult; and the shy, wary nature of this species make it more so. It does not fly with the loud wingbeats of some pigeons and makes little noise when feeding. If approached, it quietly leaves from the opposite side of the tree or sits motionless, flushing only at the last moment.

Sometimes White-headed Pigeons are seen on the ground in the rainforest but it is not known whether they are actually feeding or obtaining water. It is also possible that they are collecting grit. Unlike most fruit-eating pigeons in Australia, White-headed Pigeons digest the seed of the fruit as well as the flesh. This requires a strong, muscular gizzard containing grit to break down the hard seeds. The nest (placed in thick foliage) is usually frail, though some are more robust than the usual pigeon nest. A single chick is raised, hatching after 20 days. At the end of another 20 days, it is ready to leave the nest.

The movements of this bird are a combination of a semi-regular pattern and local nomadism. Any apparent seasonality in the appearances of the White-headed Pigeon is probably due to its close synchronisation with the fruiting cycle of its food trees. Irregularities in this schedule account for the local variation.

Topknot Pigeon

This relationship between movements of birds and ripening of fruit is noted in other rainforest pigeons. One species in which it is well documented is the striking Topknot Pigeon *Lopholaimus antarcticus*. In areas where this species demonstrates regular movements there is a fairly regular pattern of fruiting, while in others erratic development of food sources leads to sporadic occurrences of this pigeon. Large concentrations of Topknot Pigeons may form around plentiful local food supplies. In northern New South Wales there are three predictable influxes linked with the ripening of Bangalow Palm fruits, Camphor Laurel fruits and wild figs, respectively. Movements of several hundred kilometres are probably not unusual for this highly nomadic species.

The Topknot Pigeon is entirely arboreal and completely frugivorous. Any species so dependent on a particular food source, of course, is particularly vulnerable to any reduction of that source. Topknot Pigeons were once considered very common. Flocks of several thousand birds were not unusual and feeding congregations of 20,000 pigeons were reported. The clearing of rainforest has reduced their numbers, so much so that flocks now rarely exceed a few hundred birds. This decline is continuing and it is accelerated by illegal shooting. Topknot Pigeons, which are large and make good eating, are all too easily shot as they arrive at a feed tree throughout the day.

Locating of a feeding flock is not difficult. Though the pigeons themselves are usually silent, the sounds of falling seeds, excreted by roosting birds, can be heard for some distance. Flocks of Topknot Pigeons are conspicuous. Their large size and banded tail in flight are diagnostic. They make an obvious plopping noise when landing.

There are no close relatives of the Topknot Pigeon. Its displays, which are unlike those of other pigeons, is evidence of this. In its flight display a male rises to 30 metres above the canopy then angles sharply upwards for another 10 metres. At the apex of this flight he partially spreads his tail and glides downwards on half-closed wings. Upon reaching the bottom of this glide he again flies upwards with a loud clap of the wings. Another display of this species is 'neck twining'. In this the male approaches a perched female, landing so close that their bodies touch. He then leans against her and they twine their necks, holding this pose for several seconds. During the bowing display, the male erects his crest. At the bottom of each bow he gives a squeak, not unlike forcing air from a balloon. The only common call of the Topknot Pigeon is a sharp screech, but it also has a soft 'coo' that is seldom heard.

Breeding takes place in late spring with the nest situated high in a tree, sometimes 30 metres high or more. Incubation is shared by both parents. It takes about three weeks.

Distribution
Topknot Pigeon.

Brown Pigeon

All the pigeon species so far discussed are largely or entirely arboreal. They are similarly entirely frugivorous. The Brown Pigeon *Macropygia phasianella* bridges the gap between these forms and the terrestial ones that follow. It feeds close to the ground in the understorey, particularly along the edges of rainforest, for seeds and fruit. Because seeds form a large part of its diet, the Brown Pigeon must also eat small bits of gravel to break down its food in its gizzard. Birds are often seen picking-up grit on dirt roads and tracks through the rainforest. While not recorded actually feeding on the ground, these birds no doubt eat any fallen seeds they come across.

Like so many rainforest pigeons, the Brown Pigeon has decreased since the clearing of its habitat for agricultural activities. It is still common, however, largely because it has quickly added the fruit and seeds of introduced weeds to its diet, like Wild Tobacco, Lantana, Blackberry, etc. Most of these plants invade the rainforest where it has been disturbed, such as along roads, tracks, clearings and edges. This clearing and the subsequent establishment of exotic plant species has favoured the Brown Pigeon, resulting in a rapid increase after its initial decline.

The fruiting and seeding of its food plants are more regular than those of most fruit-eating pigeons. Consequently the Brown Pigeon does not have the extreme, irregular movements exhibited by some of those species. It is nomadic only on a local scale. Brown Pigeons are usually seen singly or in pairs. They do not form flocks, though concentrations will gather around good feeding sites. Their call is frequently heard, a clear 'coo-cu-woot'. They are perhaps the tamest of all rainforest pigeons — when flushed they rarely fly off very far before landing again. In flight or at roost, its brown colour and long tail separates the Brown Pigeon from all other species. Sexes differ only slightly in colour.

Elsewhere the long tail of this species and those related to it has led to the group-name of 'cuckoo-doves'. Brown Cuckoo-Dove has in fact been proposed for the Australian species. The distribution outside this country is a patchwork one; there are populations in the Philippines, Sumatra, Java, Borneo and surrounding islands, separated from the Australian birds by closely related species in eastern Indonesia and New Guinea. Within this continent, the Brown Pigeon lives in rainforest and nearby hardwood forest where fruiting trees are accessible. It extends from the tip of Cape York to south of the New South Wales border, wherever suitable habitat is available.

The flight display of the Brown Pigeon is normally performed in an open situation rather than

Opposite: An adult Topknot Pigeon *Lopholaimus antarcticus*.

Robert Edden 1983.

above the canopy. Rising into the air, the male flies with the tail partly open, wings flapping noisily. Upon reaching the top of his rise he spreads the tail fully and points it downwards, giving his body a shape suggestive of an 'inverted saucer'. He spirals slowly in descent.

Breeding peaks in spring and early summer, but may take place any month of the year. Nests may be built in almost any situation and one or sometimes two eggs are laid and incubated for 16-18 days.

Distribution Brown Pigeon.

Green-winged Pigeon

The Green-winged Pigeon *Chalcophaps indica,* or Emerald Dove as it is sometimes called, is mostly terrestial. It feeds mainly on the ground searching for fallen fruit and seeds, or securing those it can reach from low vegetation. On occasion it will feed in trees, walking along the thick branches, but it lacks the agility of its arboreal relatives.

Another characteristic of this and other ground-living pigeons is their lack of advertising flights. In its place, the Green-winged Pigeon make a prominent display on a low branch. A male swings his tail and lower body up into the air while lowering his head, then swings back the other way, repeating this 10 to 15 times. When a female appears, attracted by the display, he begins to rapidly raise and lower his body. After this, copulation takes place.

The Green-winged Pigeon is a widespread species, distributed from India and south China through Indonesia, the Philippines, eastern New Guinea and some South Pacific islands. There are two populations of this species in Australia, which differ biologically in some aspects. The east coast population inhabits rainforest and other thick vegetation, such as coastal heaths and woodlands, if food producing trees are nearby. In the northwest the

Green-winged Pigeon is primarily a resident of monsoon forest but may at times be found in quite open habitats. Apparently this bird is sedentary or only locally nomadic throughout its Australian ranges.

Breeding reaches a peak prior to the onset of the rainy season, though it can take place in any month of the year. A twig platform, 4 to 11 metres from the ground, supports the two eggs. For 14 to 16 days the parents share the incubation — the male at night and female by day. While the female sits, the male defends the territory, quickly driving away any intruding pigeons.

Sexes can be differentiated by colour. The female is generally less brilliantly coloured than her mate. Her head and neck lack his reddish tinges, and where he has a white shoulder, she is dull grey. The rump and tail of the female are brown, not black.

Green-winged Pigeons do not form flocks, usually being seen in ones or twos walking along the ground at the edge of the rainforest. They are less commonly found in the interior of the forest, far from the perimeter. If startled, these pigeons fly rapidly through the forest trees, easily weaving their way through dense vegetation. The call, a series of

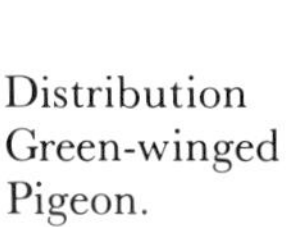

Distribution Green-winged Pigeon.

extended single notes, is repeated 10 to 12, or up to 30 times.

This species is still common in the east despite a small decline due to loss of habitat. The population of the northwest is more vulnerable. Clearing has had its usual detrimental effects but other human activities (such as mining) and the introduced animals (such as water buffalo, pigs) also present a threat in this area.

Wonga Pigeon

The monotonous call of the Wonga Pigeon *Leuco-*

Opposite: Brown Pigeon *Macropygia phasianella.*

Following pages: Green-winged Pigeon *Chalcophaps indica* (p.44) — female (left); male (right); and (p.45) Wonga Pigeon *Leucosarcia melanoleuca.*

sarcia melanoleuca is many people's first introduction to this species. The loud clear single note may be repeated 200 times or more and carries for two kilometres. At a distance it is often irritatingly just above the level of perception. Approaching a calling bird in the bush is difficult. This pigeon is very wary and long before it can be seen it has stopped calling and sneaked away. Around human habitation and recreational areas, however, this alert and shy behaviour is overcome and Wonga Pigeons become tame and tolerant of people.

The most important habitat requirements of the Wonga Pigeon are good cover and feed trees. Because they are terrestrial, these birds also need an open floor for walking. Unlike the Green-winged Pigeon, this species is not restricted to the edges of the rainforest. It also occurs in gullies of dense coastal forests and there is an isolated population in the drier inland brigalow of south-central Queensland. Within its distribution, the bird is sedentary. Not being directly dependent on the fruiting of trees, it is unlikely to have to move to find food. Feeding takes place on the ground, never in trees. It is a scavenger, selecting fallen fruit and seeds, and adding to them any small invertebrates that it uncovers.

The characteristic grey and white pattern on the front of the Wonga Pigeon is as unmistakable as its squat shape. Except when pairs come together to breed, these birds are solitary. There are no obvious close relatives and the Wonga Pigeon's courtship displays are unlike those of other pigeons. It lacks a courtship flight, this being replaced by an extraordinary advertising ceremony by the male, which takes place on the ground or on a log. Standing erect, he raises his wings, and lifts and spreads his tail. He swings his head to one side, tucking it behind one shoulder while simultaneously lowering the wings and tail. These actions are then repeated with the head swung to the opposite side. A rhythmic alternating of sides continues for some time,

Distribution
Wonga Pigeon.

each turn to the side exposing the striking pattern of the front. Bowing consists of raising the tail and lower abdomen while slightly tipping the head and breast forward.

The Wonga Pigeon has an extended breeding season. Nesting may occur at any time but the majority of activity is from October to January. The nest may be very large for a pigeon — 30cm in diameter and 9cm thick — or very small and flimsy. It is often placed in a relatively open location. An incubating bird turns its mottled underside towards an intruder as camouflage.

In part of its range, the Wonga Pigeon has become locally extinct. The usual factors, clearing and shooting, are responsible for its demise. Early naturalist John Gould previewed some of the problem when he commented on what a fine meal a Wonga Pigeon made.

Chapter 7

PARROTS

(Cacatuidae, Psittacidae; Psittaciformes)

Australia has 55 species of parrots but few inhabit the rainforests. Of those that do, three have only marginal toeholds at the northern end of Cape York Peninsula; one has three widely-separated populations along the east coast; and two are widespread but equally at home in non-rainforest habitats. Other species enter rainforests to a lesser degree but do not form a conspicuous part of the avifauna of this habitat.

Palm Cockatoo

Without question, the most striking parrot is the Palm Cockatoo *Probosciger aterrimus*. With its enormous bill, red facial skin and long crest, this bird is Australia's largest and most majestic parrot. Unfortunately very few see it in the wild, its Australian range being restricted to the tip of Cape York Peninsula and a short distance along each coast. In New Guinea, the Palm Cockatoo is widely distributed through the lowland but, because of hunting pressure for its feathers, it is scarce around human settlements. Though its Australian range is limited, it is at least common there, its preferred habitat being lowland rainforest where it borders eucalypt woodland.

This bird is easy to locate from its loud two-note call 'hweet-kweet', the first note mellow, the second shrill and rising. Other noises are made when perched by drumming on the sides of hollow trees with its bill, foot or — as one observer has noted — a stick specially chosen from a neighbouring tree. When spotted, these cockatoos may be roosting in bare branches at the top of a tall tree or flying over the canopy of the forest. In flight, the wing beats are slow and deliberate, interspersed with much gliding.

Soon after sunrise several Palm Cockatoos will congregate at a tall tree before moving off to feed. As more birds arrive, the calling increases and individuals become more active. A peculiar display is given: on the second note of its call, a bird will tip forward on the branch, hanging head down, crest and tail raised and wings outstretched. When excited the red of the face 'blushes', turning noticeably darker.

Feeding parties move into woodland adjacent to the rainforest, picking leaf buds, fruit and seeds from the trees. Birds will also feed on the ground,

Distribution
Palm Cockatoo.

particularly to obtain Pandanus fruit, a major component of their diet. About mid-morning, the cockatoos return to the rainforest where they spend the remainder of the day. Just before dusk, they retire to their roosting trees for the night.

Breeding starts in October/November. They select a tree hollow for a nest and line its floor with twigs they crush with their bills. The single white egg is incubated by the female alone for 31 to 35 days. The chick remains in the nest hollow for 14 to 16 weeks and is dependent on both parents for food for a further six weeks after having left the nest.

Recent protection of the Palm Cockatoo's rainforest habitat in reserves will help this species prevail in the face of agricultural and mining activities on Cape York Peninsula.

Two other cockatoos make occasional use of the rainforest. The Yellow-tailed Cockatoo *Calyptorhynchus funereus* resides in a range of timbered country, including rainforest. Most of its food, consisting of the larvae of wood-boring insects, seeds, fruits and

other plant material, is obtained in eucalypt forest, heathland and exotic pine plantations. The familiar Sulphur-crested Cockatoo *Cacatua galerita* may nest in rainforests, the large flocks seen in the open countryside replaced by small groups. In New Guinea, it is a wary bird of the closed forest, hunted for its feathers.

Double-eyed Fig Parrot

In contrast to the Palm Cockatoo, the diminutive Double-eyed Fig Parrot *Opopsitta diophthalma* is Australia's smallest parrot. Three isolated populations occur along eastern Australia, each with a different colour pattern on the face and distinct alternative names. The title 'double-eyed' comes from the impression given by the facial markings of some subspecies when viewed from the front of two eyes close together. Fig parrots are also called lorilets.

In a small area around Iron Range on the east coast of Cape York Peninsula is Marshall's Fig Parrot *O. d. marshalli*. Being only 130mm long, it is the smallest of the Australian forms with pronounced sexual dimorphism. Around the Atherton Tableland and surrounding rainforest areas lives the Red-browed or Blue-faced Fig Parrot *O. d. macleayana*, shown here. Differences between the sexes are principally limited to the colour of the cheeks. The largest of the subspecies is Coxen's, Blue-browed or Red-faced Fig Parrot *O. d. coxeni*, centred on the rainforest around the Queensland-New

Distribution Double-eyed Fig Parrot.

South Wales border. Sexual dimorphism is almost absent in this form. Other subspecies of the Double-eyed Fig Parrot inhabit lowland New Guinea. Females, although less brightly coloured, are heavier and bulkier than males.

Opposite: The Palm Cockatoo *Prosciger aterrimus*.

Head studies of the Double-eyed Fig Parrot *Opopsita diophthalma*. In each of the pairs the female is shown above the male. The pairs are (from left): Red-browned Fig Parrot *O.d. macleayana;* Coxon's Fig Parrot *O.d. coxeni;* Marshall's Fig Parrot *O.d. marshalli.*

The southern population of fig parrots (*O. d. coxeni*) has been seriously diminished by the removal of lowland rainforest for timber and agriculture. There have been a few recent sightings but it is now considered rare. Further shrinkage of its southern habitat may push this bird into local extinction. Other Australian populations are much healthier but individuals are not easily seen. Their small size and green plumage render these parrots hard to detect among the leaves. Small parties may be seen flying swiftly above the forest, uttering a shrill 'tseet'. Fallen, partially-eaten fruit under a tree is indicative of a favourite feeding station. The sound of falling fruit is also a clue, though the birds themselves remain quiet. As the name suggests, the preferred food is the seeds of wild figs. Some insects and their larvae are taken as supplements. Fig parrots are completely arboreal, never coming to ground.

Little study has been made of the reproductive biology of these birds in the wild. Two white eggs are laid in a hollow which has been excavated by the parents in rotten wood or in a dead limb. The length of the breeding season is not known, but active nests have been found in August and November. Only the female incubates but both parents feed the chicks.

The screeching Rainbow Lorikeet *Trichoglossus haematodus* is ubiquitous along the east coast in every forest or woodland where flowers or fruit can be

49

Robert Edden 1983

found. Lorikeets have specialised tongues to collect the nectar and pollen the flowers provide.

Eclectus Parrot

So different in coloration are the sexes of the Eclectus Parrot *Eclectus roratus* that for many years males and females were considered separate species. The bright green male has large patches of red, usually hidden by the wings; thus it is also called the Red-sided Parrot.

The Australian range of the Eclectus Parrot is limited to the Iron Range district of Cape York Peninsula. Widespread through New Guinea, the Solomon Islands and the Moluccas, this species was known in these areas long before it was discovered in Australia. In 1913, William McLennan obtained the first specimens from far north Queensland.

The Eclectus Parrot may be noisy and conspicuous but it is a wary inhabitant of rainforest, gallery forest and adjacent woodland. Within its restricted Australian distribution it is common, its raucous calls being a feature of this area. From roosts in the rainforest canopy, flocks move into the woodland to feed on fruit, nuts, seeds and blossoms. When they return in the evening, there is considerable racket and commotion before they settle down for the night.

The nest is placed in the bottom of a tree hollow, often 10 metres below its entrance. The male feeds the female as she incubates the eggs for 26 days, after which both care for the young. There are many records of additional birds, sometimes up to eight of both sexes, also gathering at the nest. The significance of this is unknown, but as Eclectus Parrots frequently raise two broods in a season, it has been speculated that these birds may be older offspring which remain with the parents for some time after leaving the nest.

Distribution
Eclectus Parrot.

Double-eyed Fig Parrot (Atherton District) *Opopsitta diophthalma* — female (above); male (below).

A male Red-cheeked Parrot *Geoffroyus geoffroyi*

Red-cheeked Parrot

The Red-cheeked Parrot *Geoffroyus geoffroyi* lives in the same part of Cape York as the Eclectus Parrot. They were discovered at the same time by William McLennan and both are certainly more widespread through New Guinea and parts of eastern Indonesia. The Red-cheeked Parrot, however, seems more closely associated with the rainforest than is the Eclectus Parrot, and it rarely moves far from it into the woodland. Nevertheless there have been reports of the Red-cheeked Parrot feeding in nearby mangroves.

Distribution
Red Cheeked
Parrot.

The red cheek from which this species receives its name is found only in the adult male. Females have a brown head; in immatures it is green. This species' green body plumage blends well with the upper foliage. The shrieking 'aank aank aank' call is heard more often than the bird is seen.

Not much is known about the Red-cheeked Parrot in Australia. Breeding has been recorded

Robert Felden 1984

from August to December but evidence suggests the season may be longer. Pairs are often at very different stages in the breeding cycle during this period. Two to four eggs are laid in a hollow. Young birds stay with their parents for some time, possibly for as long as two years.

King Parrot

The brightly coloured King Parrot *Alisteris scapularis* is a well-known coastal species. It is usually encountered away from the rainforest because it becomes tame where fed, or even around human habitation to which it can become adjusted. Elsewhere it is wary and difficult to approach. In rainforest, King Parrots spend much of their time in the canopy, hidden by the foliage. Their calls are often the first indication of their presence. Nesting ordinarily takes place in a hollow in a eucalypt, thus it is in the thick wet forests and more open coastal woodlands that this species is most common. This parrot also visits parks and gardens where trees are in flower. Orchards and grain crops attract raiding birds in search of food.

King Parrots are usually observed in pairs or small groups, the brilliant red and green adult male being the most obvious. Most of the birds seen, however, are the less colourful females and immature males. Young males retain a female-like plumage for 16 months. For the next 14 to 15 months they slowly acquire the coloration of the adult, though they are physiologically capable of breeding while still in immature plumage.

At the commencement of breeding, a courting pair perform a spirited ceremony. The male fluffs up his head feathers, sleeks down the remainder of his plumage, flicks his wings and calls to the female. She responds with a similar display. She bobs her head until he ceremonially feeds her.

A hollow limb or tree trunk is selected for nesting. The bottom of the nest chamber is usually some distance below the entrance. Some as deep as 10 metres have been recorded. Three to five white eggs are laid on the chamber floor. The female, without any assistance from the male, sits on the eggs for 20 days. During this period her mate periodically brings food. Five weeks after hatching, the young leave the nest, remaining with their parents to form the family parties usually seen together.

After the breeding season, immatures form wandering flocks of 20 to 30 birds. There is also evidence that King Parrots have an altitudinal migration between seasons, moving to lower elevations in winter and returning in summer. In the Atherton Tableland, this species has been reported at altitudes of 1625 metres.

King Parrots eat a range of vegetable matter: fruit, seeds, blossoms, leaf buds and nectar. They forage in the foliage for growing items or on the ground for spilt grain and fallen seeds. If disturbed, they fly with a heavy wingbeat, though they are surprisingly swift and agile. An alarmed shriek is given in flight. Other calls include a bell-like piping note and a shrill 'crassak crassak'.

The nearest relatives of the King Parrot of eastern Australia are two similar species in New Guinea.

Distribution King Parrot.

Crimson Rosella

One of the most familiar parrots of eastern Australia is the Crimson Rosella *Platycercus elegans*. It is common to abundant throughout its range, favoured habitats being rainforests and the humid eucalypt forests of the coast and adjacent ranges. When feeding or roosting in the tops of trees, the Crimson Rosella, despite its gaudy plumage, may be difficult to see. More often it is observed in small flocks or family parties foraging on the roadside. Not a shy species, it may be closely approached — a tameness that has led to many birds being struck by cars. Casualties include a high percentage of immature birds, who are even more incautious than the adults. These younger birds are easily distinguished by their green rather than red plumage. Roadside feeding parties may consist of a family group of parents and

Following pages: Eclectus Parrot *Eclectus roratus* (p.54) — female (left); male (right) and (p.55) adult pair of Crimson Rosellas *Platycercus elegans*.

Robert Felden 1984

Robert Fdden 1983

offspring, or of a roaming flock of immatures —
sometimes up to 50 or more — which gather in the
post-breeding season.

Breeding commences with the arrival of September and continues until January. Males display
to their mates by puffing out their chest, drooping
their wings and wagging the fanned tail from side to
side, accompanied by continual chattering. Rosellas
apparently mate for life. They do not seem to defend
a territory with the possible exception of a small area
directly around the nest hollow. This hollow is
usually 8 to 30 metres from the ground, in a dead
limb or tall eucalypt. The female incubates alone
but she leaves the eggs twice a day to find food to
supplement that brought by her mate. After 19 days
the eggs hatch. When the chicks emerge from the
nest after five weeks, they have the characteristic
green feathers. For another month the family
remains together, after which the offspring leave to
join other immatures. Flocks of young birds stay

Distribution
Crimson Rosella
(red forms only).

together for 15 months, during which period they
start to acquire the red feathers of adults. By the end,
few green feathers are still present in such flocks.

Non-breeding adults disperse after the departure of the young, moving into more open areas such
as parks, suburban gardens, roadside trees and
wooded paddocks. Crimson Rosellas are at home
even in the alpine country; they have been observed
at 1900 metres on Mt Kosiusko. In such areas they
make an attractive sight as they search for seeds on
the surface of fallen snow. A variety of food is consumed. Seeds, fruit, blossoms, buds and other plant
products are augmented with insects and their larvae. Cultivated fruit orchards are raided, a practice
that makes this parrot unpopular in some areas.

Crimson Rosellas have a range of calls. When
alarmed, they give a shrill screech. A clear bell-like
series of two or three notes and a musical chatter
seem to help birds maintain contact while feeding or
in flight.

There is an interesting story surrounding the
origin of the name 'rosella'. Early settlers found this
bird at Rosehill (now Parramatta), New South
Wales, and called it the Rosehill Parrot. This was
eventually abbreviated to 'rosehillers' and in time
was corrupted to 'rosellas'.

The Crimson Rosella has an isolated population of smaller darker birds in northeastern
Queensland which may lack a green immature
plumage. Other subspecies in the east and southeast
— and one on Kangaroo Island — are also red. Very
closely related forms occur in the Mt Lofty Ranges
district (the orange Adelaide Rosella) and along the
Murray-Murrimbidgee Rivers system (the Yellow
Rosella). A larger species, the Green Rosella *P.
caledonicus,* replaces the Crimson Rosella in Tasmania; here it inhabits both open country and cool
temperate rainforest. All these rosellas are very
similar in their habits.

Chapter 8

CUCKOOS

(Cuculidae; Cuculiformes)

Many cuckoos have evolved an approach to breeding that differs from the conventional behaviour of most birds. Instead of constructing a nest, incubating the eggs and rearing the young themselves, these birds practise 'brood parasitism'. This involves laying their eggs in the nests of other species (the host), leaving that duped bird to perform the duties of raising the young. When a nest is unattended, the female flies in and quickly deposits an egg. Some cuckoos remove an egg of the foster parents at the same time.

The eggs of some cuckoos are quite different in colour from those of the host; in others the resemblance in colour and pattern are impressive. Cuckoos' eggs require a few days shorter incubation period than those of their foster parents. A newly hatched cuckoo grows quickly and soon outgrows the chicks of the host. It may expel the other eggs or nestmates, or simply outcompete them for the food brought by the parents.

The host species do not seem to recognise the sudden appearance of a foreign egg or the presence of a large alien nestling; the open-mouthed begging squawks of the chick compel the parents to continue their efforts of providing food. In most situations, the rapidly growing cuckoo is soon far larger than its unwitting hosts. This feeding continues well after the cuckoo has left the nest. Many birds appear to recognise the now nearly full-sized bird as a potential threat and attack it. Its loud protests, however, suppress their aggressive urges, replacing them with the impulse to also contribute food. Even birds other than the foster parents may be seen feeding the young.

Australia has 14 species of cuckoo, all but one being brood parasites. The exception is the Pheasant Coucal *Centropus phasianinus,* which is quite different in appearance and behaviour, performing all the tasks of raising its own family. Several of the parasitic species are migratory. They arrive in their breeding areas in spring and start to search for the nests of potential hosts. Loud persistent calls announce their presence, often continuing throughout the night until their shrillness and monotony become annoying.

Hairy caterpillars are a favourite food of some species while others show a strong preference for fruit. Many of the Australian cuckoos enter rainforest in search of food or to hunt for their host species.

Species which use rainforest include the Oriental Cuckoo *Cuculus saturatus*, Brush Cuckoo *C. va-*

Chestnut-breasted Cuckoo *Cuculus castaneiventris.*

riolosus, Fan-tailed Cuckoo *C. pyrrhophanus,* Chestnut-breasted Cuckoo *C. castaneiventris,* Shining Bronze Cuckoo *Chrysococcyx lucidus,* Koel *Eudynamys scolopacea* and Channel-billed Cuckoo *Scythrops novaehollandiae.* Of these, perhaps the most dependent on rainforest — and certainly the least known — is the Chestnut-breasted Cuckoo. Though widespread in New Guinea, it is restricted to Cape York Peninsula in Australia. It resembles the better-known and more widely distributed Fan-tailed Cuckoo, but it is darker with its underparts a rich chestnut colour. Like that species, it has a descending trill and shares a liking for hairy caterpillars. The Fan-tailed Cuckoo is known to parasitise at least 50 other species. But any Australian hosts of the Chestnut-breasted Cuckoo are as yet unknown because no definite nests or eggs have ever been found. There is even debate upon whether this species breeds in this country. Some ornithologists consider it a non-breeding visitor from New Guinea. Others believe it is sedentary. Because the Fan-tailed Cuckoo also occurs in the same areas, it would probably be very difficult to separate the eggs anyway. Similarities in voice further contribute to the confusion. Any confirmation will require new observations, particularly of the Chestnut-breasted Cuckoo actually parasitising a nest.

In southeastern Australia, the cuckoo of the rainforest is the Brush Cuckoo. Its shrill call, a series of descending notes, is repeated incessantly. Another call phrase, which, becomes a feature of rainforests in the spring, rises quickly to an almost hysterical finish.

The large Koel is also an extremely vocal bird at times. Several calls including a 'koeel' or ascend-

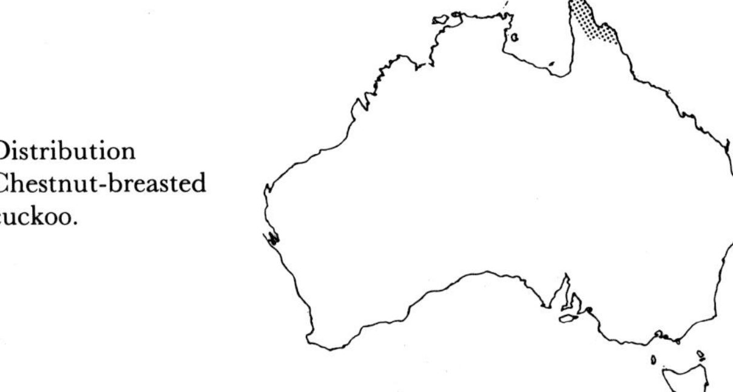

Distribution
Chestnut-breasted
cuckoo.

ing 'weeir weeir weeir' are given continually, day and night. This can make it an unwanted and irritating guest if it is within earshot. Fruit rather than caterpillars is the major constituent of its diet. The nests of currawongs, miners, wattlebirds and other large birds are victimised by this cuckoo.

Chapter 9

FROGMOUTHS

(Podargidae; Caprimulgiformes)

The Tawny Frogmouth *Podargus strigoides* is well-known to many, if not often seen. It belongs to a small family of strange birds which receive their names from the broad flattened bill that gives their faces a somewhat frog-like appearance. Three species are found in Australia: the Tawny Frogmouth of open country; the Marbled Frogmouth *P. ocellatus* of the rainforest; and the Papuan Frogmouth *P. papuensis* which shares bits of both. The latter is the biggest of these birds, growing to more than half-a-metre in length.

The Papuan Frogmouth is similar in behaviour to the Tawny Frogmouth. During the day they roost quietly on a branch or stump, often in communal gatherings of several birds. If approached, they stretch their bodies upwards and point their bills towards the sky. Such a posture and the camouflage of its mottled plumage make it resemble a broken limb. At night these birds become active and begin their search for food. Papuan Frogmouths spend the day roosting in the edge of the rainforest but move into adjacent open woodland to feed.

Frogmouths are often mistakenly believed to be owls but there are obvious differences in the way

talons. In contrast, frogmouths have weak feet and blunt claws; their feet, unlike those of the owls, are incapable of grasping and killing prey. Instead, a frogmouth pounces on its victims from a branch, snapping them up and crushing them in its wide bill. The item is either eaten on the ground or carried back to a perch where it is beaten against a branch to complete the kill or to soften it up. A range of small animals are eaten. The wide, flat, bony roof of the mouth may help protect the bird against the large spiny insects which form a large part of its diet. This bill is also used to make clacking noises at intruders which approach the nest too closely.

Distribution
Papuan Frogmouth.

Distribution
Marbled Frogmouth.

they feed and in the structure of the bill and feet. Owls have much narrower but more hooked bills, and their strong feet are armed with heavy sharp

Smaller birds seem unable to make the distinction between owls and frogmouths. When they find a roosting owl during the day it is a common practice for them to gather round, calling and harassing in an attempt to drive it away. They similarly mob the comparatively harmless frogmouths.

Papuan Frogmouths build a small platform of twigs, lined with green leaves, which they place in a horizontal fork or crotch between a limb and trunk. Both sexes share the duties of raising the young, but the process is so prolonged that only one brood is reared each season. This usually consists of a single white egg, sometimes two. Frogmouth chicks are bundles of fluffy down.

Robert Edden 1983

The Papuan and Marbled Frogmouths also occur together in New Guinea, the latter reaching the Solomon Islands. In far north Cape York Peninsula, the three Australian species can be found together. The large size and mottled plumage of the Papuan Frogmouth are distinctive but in a spotlight the best character is often the red eyes. The plumage of the Marbled Frogmouth is not unlike that of the Papuan, though its eyes are orange or yellow. This species has two Australian forms. Southern birds are about the size of the Tawny Frogmouth but those of Cape York Peninsula are very small, only two-thirds the length of a Papuan Frogmouth, a substantial part of which is accounted for by the proportionally longer tail. At the base of the upper bill is a tuft of long erect feathers from which is derived the alternative name, Plumed Frogmouth.

Voice is another clue to identification. The Marbled Frogmouth has a soft 'koor-loo' repeated over and over while the Papuan Frogmouth gives a series of 'ooms' and a bubbling 'hoo-hoo-hoo' descending as it progresses. A continuing 'ooo-ooo-ooo' is characteristic of the Tawny Frogmouth.

Unlike the other two species, the Marbled Frogmouth rarely leaves the rainforest. This and its nocturnal habits has made it a difficult bird to study. Very little is known about it and it is considered rare. The small Cape York birds are apparently the more common of the two forms, according to records of the early naturalists. The southern population was, until recently, known only from a handful of records. Though by using tape recordings to which the frogmouths respond vocally, it has been found that both numbers of birds and extent of distribution are greater than previously believed. There is only a small amount of information on the breeding of the northern bird and nothing on the southern one. No nest or egg of the latter has ever been found. Nest materials of the Marbled Frogmouth on Cape York are more interwoven than those used by the Papuan Frogmouth. This frogmouth is not known to adopt a 'broken limb' posture either, nor is it known to form communal roosts.

Opposite: Marbled Frogmouth *Podargus ocellatus*.

Chapter 10

OWLS

(Strigidae, Tytonidae; Strigiformes)

Owls have long captured man's imagination and often inspired his fears. The forward facing eyes give them a more human appearance than other birds. Their association with the night has also set them apart. This is not only the time that they become active, but they do so as masters of darkness. Several features of owls' bodies contribute to their remarkable abilities. The large eyes are extraordinarily sensitive. They can detect objects at an illumination of between one-tenth and one-hundredth of the intensity required by man. Their head is mounted on a highly flexible neck allowing rotation of 270 degrees. So without moving, an owl commands a view of its entire surroundings. Soft fluffy plumage and the structure of the leading edge of its wings permit silent flight as the owl dives on an unsupecting victim. Even in the complete absence of light, owls are capable of searching out and attacking prey. The ears are slightly offset at the sides of the head. Such an arrangement causes sound waves to strike at minutely different times; intervals as short as 0.00003 seconds can be perceived by the owl. This sensitive hearing can guide it to the source. Being birds of the night, owls rely on loud cries to communicate and maintain contact. Often eerie booms or loud shrieks, these nocturnal screams have helped build owls' malevolent and mysterious folk image.

There are two major divisions of owls; four of each are found in mainland Australia. Hawk owls, including species such as the familiar Boobook Owl or Morepork *Ninox novaeseelandiae*, have the typical appearance familiar to most people. The other group is the 'monkey-faced' owls, so-named for the definite heart-shaped facial disks. One species of each family occurs in the rainforests of Australia. Neither is common and their patchy distributions and nocturnal habits have meant that much of their natural history has yet to be adequately studied.

Rufous Owl

In rainforests of northern Australia and New Guin-

ea, the hawk owls are represented by the large Rufous Owl *Ninox rufa*. Most of its daylight hours are spent roosting within the dense vegetation of the forest. As night approaches, this bird wakes and prepares to hunt for food by moving into the more open habitats fringing the rainforest. A Rufous Owl may travel up to four kilometres from its roost in a night in search for food. Males and females roost together during the day but go forth to feed separately. Like all owls, they are strictly predatory. Mammals, including flying foxes, large stick insects and a high percentage of birds are taken. Prey up to the size of the Brush Turkey are attacked with the powerful feet and talons. In order to better grip the struggling animal, one of the front toes of owls is reversible giving them a wider grasp. Undigested bones and fur are coughed up as a compact pellet. A buildup of pellets and droppings at the base of a tree mark a favourite feeding post.

Rufous Owls announce their territories with a drawn out 'whooo-hoo', which is given with greatest frequency early in the breeding season. This starts in winter and continues into spring. Territories are permanently maintained by a pair of owls which mate for life. A hollow in a tree serves as the nest and may be reused for several successive seasons. Before it is ready for use, the hollow must be cleared of the

Distribution
Rufous Owl.

Opposite: Rufous Owl *Ninox rufa.*

droppings of the young and half-eaten remains of prey accumulated during the previous season. By the time this is underway, the nesting pair have become silent, rarely giving the territorial calls heard earlier. Rufous Owls, like other species, lay white eggs. The usual clutch is two and only one brood is raised per season. The eggs are placed on a layer of wood debris at the bottom of the hollow and incubated by the female for five weeks. The male, usually a shy bird, will at this time defend the nest vigorously by diving at intruders. His powerful talons are an effective deterrent.

Sooty Owl

Even bigger than the Rufous Owl is the Sooty Owl *Tyto tenebricosa,* particularly the female. She is considerably larger than the male with much more substantial feet and claws. This discrepancy between the sexes may allow them to specialise at different sized prey. Sooty Owls kill a larger proportion of mammals than do Rufous Owls, and complete their diet with birds. No insects, an im-

Distribution Sooty Owl.

portant component of the diet of the previous species, appear to be eaten. The Sooty Owl is not only one of the largest of the Australian 'monkey-faced' owls, it is also the most adapted for life in the rainforest. It does not leave the confines of this habitat to feed as it is able to manoeuver and hunt within the dense vegetation. Unsuspecting victims may be attacked whether on branches of the canopy or on the forest floor.

Like the Rufous Owl, the Sooty Owl is sedentary and guards a territory throughout the year. It has an interesting repertoire of calls but the most

noted, and certainly the most disconcerting at close range, is a loud descending scream. It is known as the 'falling bomb' call among ornithologists.

Sooty Owl *Tyto tenebricosa.*

During courtship, the mated pair engage in twittering duets around the nest hollow. Preparation of the hollow is similar to that of the Rufous Owl. The female incubates the two eggs for six weeks. She remains on the nest without a break, depending on her mate to bring food. The young do not fledge for another six weeks after hatching.

Sooty Owls occur in both Australia and New Guinea. The Australian birds are found in two quite separated ranges. One is along the southeastern coast and adjacent ranges, the other in a small section of northeastern Queensland. It has recently been suggested that the latter bird may be a distinct species. They are smaller than either those to the south or in New Guinea, and show small but consistent differences in plumage. The feet and claws do not exhibit the obvious sexual dimorphism of the larger birds. Whether this is also reflected in the foraging behaviour of this population is not known.

Chapter 11

KINGFISHERS

(Alcedinidae; Coraciiformes)

The best known of Australia's ten kingfishers is the Laughing Kookaburra *Dacelo novaeguineae*. It occasionally enters rainforest but is much more common in other habitats. The two species which spend most of their time in rainforest are brightly coloured natives of Queensland and New Guinea.

White-tailed Kingfisher

The attractive White-tailed Kingfisher *Tanysiptera sylvia* belongs to a group of similar long-tailed species known collectively as racquet-tailed or paradise kingfishers. This species has the accurate but clumsy alternative name, Buff-breasted Paradise Kingfisher. The tail makes up more than half of the total length.

Each year these birds migrate between Australia and New Guinea. Their arrival in October/November is heralded by their loud persistent calls as they establish their territories. This involves a constant and high-pitched trill throughout the day during the breeding season. The other call of this kingfisher is a 'chuga chuga chuga', repeated six or seven times. The major requirement for this species is the presence of active termite mounds on the floor of dense lowland rainforest. Once mated, the adult pair begin to excavate a nest hollow in the side of a termite mound. When completed, the tunnel will be 130mm high and 150mm in diameter. Digging the tunnel requires three to four weeks. The entrance is usually within 300mm to 350mm off the ground, though may occasionally be as low as 150mm. The kingfishers grow much more silent once nesting begins.

A week after the nest hollow is ready, the female lays one to three round white eggs. These hatch with remarkable regularity, generally between the last week in December and the second week in January. The rapidly growing chicks remain in the nest for 24 days. During this time they are noisy and aggressive, fighting among themselves and causing much of their food to be dropped to the floor of the chamber. A kingfisher's nest is not a pleasant area with its combination of droppings, decaying food and fly maggots. Nor do the young look as cute as most baby birds. In most birds, the feathers emerge covered in pulpy sheaths, but soon break through, leaving some remains of the sheaths at their bases. On baby kingfishers, however, the sheaths keep growing with the feathers for some time, giving the impression that the birds are covered in spines.

Distribution White-tailed Kingfisher.

Adult kingfishers too lose some of their attractiveness as their plumage becomes soiled and their tails abraded from entering and leaving the nest. Immatures are duller with dark bills, mottled breasts and incomplete tails.

In March/April, the White-tailed Kingfishers depart for New Guinea, the young birds leaving later than adults. Some individuals have been seen in Australia during the winter, suggesting that part of the population may remain through the year.

Yellow-billed Kingfisher

The Yellow-billed Kingfisher *Halcyon torotoro* is more

Following pages: Adult pair of White-tailed Kingfishers *Tanysiptera sylvia* (p.66); and (p.67) Yellow-billed Kingfishers *Halcyon torotoro* — female (above) and male.

Robert Fdden 1983

Robert Eddon 1983

restricted in its Australian range. Its remote locality and its choice of thick habitat have prevented much study being made of this species. Little is known of

Distribution
Little Kingfisher
and (darker area only)
Yellow-billed
Kingfisher.

its biology. Females largely resemble the males, but the orange on their front is paler and the top of their crown is black. For most of the year this kingfisher is solitary and quiet. During the breeding season, however, (November to January, though possibly longer) it becomes more vocal. A loud descending trill is heard much more often than the bird is ever seen.

The Yellow-billed Kingfisher builds its nest in arboreal termite mounds or in hollow limbs 5metres to 15metres from the ground. It scoops out a tunnel of 100mm to 300mm long opening into a large chamber. Three to four white eggs are laid but few other aspects of its breeding biology have been recorded. It is less aggressive in defense of the nest than the White-tailed Kingfisher. Its parents may even desert the eggs if disturbed too much.

This species is most often found around the edge of rainforest and in fringe vegetation such as woodland, monsoon forest and mangroves. It is generally considered sedentary but fluctuations in numbers may suggest a degree of migration. In New Guinea, this bird occupies lowland rainforest. Its place is taken at higher altitudes by another species which is virtually identical in appearance.

Little Kingfisher

The smallest of Australia's kingfishers, the Little Kingfisher *Alcedo pusillus,* also enters rainforest. Unlike the other two species it is not an occupant of the forest centre. Instead it feeds along streams, diving for small aquatic animals. It is seldom found far from water in rainforest. The Little Kingfisher lives in other habitats where there are thickly vegetated water edges: mangroves, for example, are a favoured locality.

Chapter 12

PITTAS

(Pittidae; Passeriformes)

The Noisy or Buff-breasted Pitta *Pitta versicolor* is one of Australia's most brightly coloured birds. It has bold patches of contrasting colour, some of which are iridescent — particularly the blue on the bend of the wing. This gaudy appearance is characteristic of all 28 species of pittas in the Old World and has earned them the group name of 'jewel thrushes'. Yet despite its striking plumage, the Noisy Pitta is one of the most difficult birds to see in the rainforest. The dim light of the forest seems to mute its colours, making it inconspicuous at any distance, even when in full view. It is usually shy too, and makes considerable efforts to remain concealed.

The Noisy Pitta is heard much more frequently than it is seen. Its loud, ringing call — a whistling three-note 'walk-to-walk' or 'want-to-watch' — carries for a considerable distance. While calling, they may frequently be perched as high as ten metres up a tree. Pittas respond quickly to an imitation of their call, even a crude one. They silently move towards the caller, either hopping across the ground or flying quickly from perch to perch, pausing every so often to answer with their own call. This is often the only way to get a good look at this bird.

The Noisy Pitta reveals a large patch of white in each wing when it flies, though this is concealed while the bird is at rest. On the ground, it stands upright on 40mm-long legs, hopping like a thrush. Their distinctively squat look is caused by the very short tail. Pittas in the southern part of their range are larger (2000mm) than those from the Iron Range area of Cape York Peninsula (175mm).

Noisy Pittas occur along the eastern coast of Australia, south to Port Macquarie, New South Wales (though wanderers have been recorded as far south as Sydney). They prefer rainforest but will occasionally enter wet coastal forests. This species is also found in southern New Guinea and in fact many of the Australian birds migrate northwards in winter, even crossing the Torres Strait. Other populations here seem to remain resident throughout the year. The annual movements of the Noisy Pitta are not well understood. Recognition of north-south migration is further confused by apparent shifts in altitude through the year: pittas from highland areas appear to move to lower elevations during winter.

Noisy Pittas start to breed in October, or sometimes as early as August. The nest is a large bulky dome — approximately 300mm x 300mm x 200mm — with a side entrance. It is constructed of sticks, leaves, bark and roots. The inside of the nest chamber is then lined with decayed wood and debris from the forest floor. Leading to the side entrance is a stage or platform 150mm long, built of animal dung. Three to four, sometimes five, spotted eggs are

A Noisy Pitta's anvil, surrounded by the shells
of its victims, in the Iluka Rainforest.

Following pages: Noisy Pitta *Pitta versicolor* (p.70); and (p.71) Blue-breasted Pitta *Pitta erythrogaster*.

Robert Edden 1983.

laid. Breeding usually finishes in January though it may extend to March.

Pittas use their bills to dig through the leaf litter for a range of invertebrates and small fruits, though the food for which they are most noted is snails. These they carry to a flat rock known as an 'anvil', where, holding the snail in their bill, they strike it

Distribution
Noisy Pitta.

against the rock until its shell breaks. Discarded snail shells litter the edges of a frequently employed anvil which may be worn smooth by use. Other objects may substitute for a flat rock, even a piece of wood. Dave Cowen, of Iluka, observed a pitta that had chosen a discarded piece of green bottle for its anvil.

The name 'pitta' is of Indian origin. Noisy Pittas were also called 'Dragoonbirds' because their colour pattern was as bright as the uniform of a dragoon, or cavalryman.

Another species can be found in the rainforests of northern Cape York Peninsula. This is the Blue-breasted Pitta *P. erythrogaster*, also called the Red-bellied Pitta, a particularly apt name. This species is every bit as brilliant as the Noisy Pitta but has a strikingly different colour combination. In much of its behaviour, the Blue-breasted Pitta resembles its more widespread relative. It migrates to New Guinea in the dry season after breeding. This species' mournful whistle is quite unlike the call of the Noisy Pitta.

Distribution
Rainbow Pitta
and (darker area only)
Blue-breasted Pitta.

A third species is the Rainbow Pitta *P. iris*, which resides in the monsoon rainforests, riverine scrubs and mangroves of the northern edge of Australia. It is sedentary within this narrow range. This species looks and behaves like a small Noisy Pitta except it is black where that bird is buff.

Chapter 13

LYREBIRDS

(Menuridae; Passeriformes)

Australia is a land containing many remarkable birds but few as remarkable as the lyrebirds. The only two species, confined to eastern Australia, are known for their spectacular courtship displays and outstanding vocal mimicry.

Superb Lyrebird

The best-known and most widespread is the Superb Lyrebird *Menura novaehollandiae*. From just south of Brisbane to near Melbourne, this bird inhabits rainforest, wet eucalypt forest, densely vegetated gullies and adjacent woodlands. It is also making itself at home in man-made habitats, such as gardens and plantations of exotic pines, when they are close to its native habitats. The Superb Lyrebird is usually shy and is quick to flee if approached. Most people get little more than a fleeting glance as one dashes across a road or darts through an open spot in the protective scrub. In a few areas, however, this bird has become tame, allowing excellent views as it goes about its business; Sherbrooke Forest in Victoria is the best known of these.

Hearing a lyrebird is no such problem. They become extremely vocal during the breeding season, with loud ringing cries confirming their presence. The Superb Lyrebird's abilities as a mimic are well-known. It incorporates the songs of other birds into its repertoire, interspersing them with calls of its own. Each species is heard for several phrases and then the vocalist passes on to the next. Even flocks of calling birds are imitated with the lyrebird simultaneously performing all parts and, if they are in flight, adding the wingbeats as well. Occasionally man-made sounds are included — cars, whistles, machinery. One bird learned the whistle that a farmer used to call his dogs — much to the distress of the dogs who continually answered the wrong 'master'. In the New England Tableland a captive bird was released after having acquired the musical scales from his owner who played the flute. Gradually lyrebirds in adjoining territories picked up the notes from him and incorporated them in their own songs. Snippets of the scales played by a flute may still be heard in this district. This demonstrates that lyrebirds learn new songs from other species and from each other; young birds seem to adopt much of their selection from those of older neighbouring birds.

Superb Lyrebirds sing as they scratch through leaf litter in search of food but the best performances take place when the male is courting. Starting in mid-winter, the male constructs a display mound by raking away all the vegetation with his large, powerful feet. Eventually he removes all unwanted material, leaving a bare patch of soil pushed into a mound about a metre wide and 150mm high. He may build several mounds within his territory. These serve as the focal points for his displays to attract mates while warning other males of his territorial boundaries.

Displays consist of spreading the tail over the back, accompanied by loud bouts of singing. The male Superb Lyrebird has an elaborately adorned tail with specialised feathers. Two outer feathers, known as lyrates, are curved into the shape of the ancient musical instrument, the lyre. Each is black and reddish above with indented sections or notches. The rest of the tail consists of two thin, down-curved plumes and 12 lacy feathers (the 'strings' of the lyre). The upper surfaces are dusky brown but the surfaces beneath each feather have a silvery sheen. During his ceremonies the tail, which is normally carried horizontally behind, is thrown forward over the back and spread to expose the metallic undersurfaces. As he sings, the male turns in slow circles on his mound, shaking the spread tail. This is the aspect shown on the reverse side of the Australian ten cent coin. Males are larger than females, reaching a length of a metre of which the tail comprises over half. The tail feathers of the female lack the ornamental structure.

As the display continues, females are attracted to the mound. At the peak of the ceremony, the male may dance backwards and forwards with the tail still spread. He mates briefly with as many females as he can, who then leave to set up breeding territories of their own, established around the perimeter of the male's. From year to year the same male and group of females appear to associate. The male renders no assistance in raising the family or building the nest. Each female builds her own large dome of sticks, bark and other debris. This is set on the ground, in a crevice, a broken stump, or a tree fern. She lines the interior with moss, rootlets and a layer of her own feathers. Lyrebirds lay only a single large,

Superb Lyrebird nest.

purplish-brown blotched egg, which the female incubates for six weeks. A further six weeks pass before the fluffy chick emerges from the nest. Parties of females and immatures gather in the non-breeding season. The singing of the male begins to dissipate with the onset of spring as his old tail feathers are gradually replaced with new ones, a moult that is completed by January.

When not engaged in breeding activities, much of the lyrebird's daily routine involves finding food. Evidence of these searches are disturbed patches of soil covered with deep scratches from their feet. Rotten logs are often torn apart in the same manner. A range of invertebrates and small vertebrates living in the litter are eaten. At night, lyrebirds roost in trees. They are not agile or efficient fliers and climb the tree by making short half-jumps from branch to branch. Likewise when startled on the ground, they give a loud shriek and flee in a similar flurry of leaps and bounds.

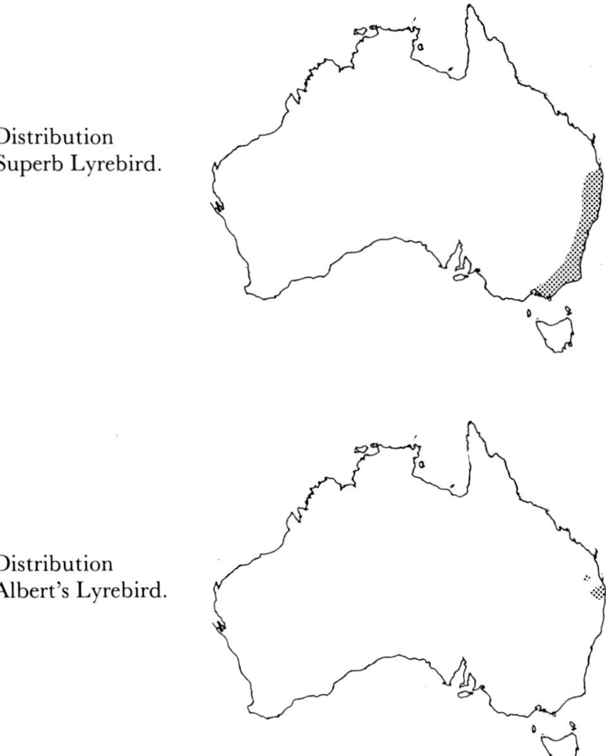

Distribution Superb Lyrebird.

Distribution Albert's Lyrebird.

Albert's Lyrebird

The Superb Lyrebird is not endangered and it has even been introduced to Tasmania, in some areas of which it has become successfully established. The other species, the Albert's Lyrebird *M. alberti* (named for Prince Albert) is far more restricted in its small range and certainly not as common. A resident of subtropical rainforest, the Albert's Lyrebird is a shyer bird than its relative, which makes it hard to spot in its dense habitat, and as a result it has been

Opposite: A male Superb Lyrebird *Menura novaehollandiae.*

Robert Edden 1983

An adult male Albert's Lyrebird *Menura alberti.*

less studied than the Superb Lyrebird. Slightly smaller, the Albert's Lyrebird is more rufous and lacks the elaborate tail feathers. In display it uses a platform of flattened vines and bush, trampled down with its feet, rather than a bare dirt mound. Division of nesting duties, egg and nest, and feeding behaviour resemble those of the Superb Lyrebird.

Relationships of the lyrebirds are obscure. They have long been considered allied to the scrub-birds and anatomical features support this. To what other group of birds they are close is a subject of active debate. Some scientists find similarities with South American ground birds. Others consider that aspects of the natural history and some biochemical studies point to a closer relationship with the bowerbirds.

Chapter 14

RUFOUS SCRUB-BIRD

(Atrichornithidae; Passeriformes)

Birdwatchers find the Rufous Scrub-bird *Atrichornis rufescens* a challenge because of its restricted range, local occurrence and elusive nature. Scientists find it a puzzle because of its enigmatic relationships and anatomical peculiarities. This bird's loud ringing voice is another of its notable features.

Located in a small area from the Border Ranges south to the Barrington Tops, New South Wales, the Rufous Scrub-bird was once considered rare. It was believed to be restricted to highland stands of Antarctic beech scattered through this zone. Recent studies have provided a much clearer picture of the scrub-bird's habitat needs. It is often, but not always, associated with beech forest. Its major requirements are a combination of wetness and thick vegetation within a metre or so off the ground. Where such conditions exist, this species may be found whether it is dry sclerophyll, logged wet forest or beech forest. Edges of beech forest often have these suitable conditions; the interior of this habitat, however, is usually too open. It is also known that the Rufous Scrub-bird was originally found at lower altitudes. Habitat destruction has eliminated these populations leaving only those in scattered highland patches. Fortunately most of the major concentrations of this species are now protected in national parks through recent reallocation of land. Its future is now brighter than it has been for many years.

It is not difficult to locate the territories of Rufous Scrub-birds, provided one is in the proper habitat at the right time of the year. Males become exceedingly noisy during the breeding season, advertising their territories with resonant calls that are far louder than would be expected for their size. Their varied repertoire includes a loud 'chip chip chip' call given in series and a single penetrating note. Scrub-birds are also among the most accomplished mimics of the bird world, rivalling the larger lyrebirds. If startled they have a harsh alarm call;

this is the only vocalisation the female is known to make.

A singing male may be within two metres of the observer and yet remain completely concealed by the vegetation. Attempts to move closer to the bird are usually unsuccessful; without having given any evidence of its departure, the bird will begin calling again at a different location. This species can travel quickly and quietly through the thickest scrub. It is a very poor flier and will only take wing if under extreme pressure. Even then, after a short fluttering flight, it drops back into the undergrowth. The lucky observer may see the male in full song, head thrown back and mouth wide open. Usually, however, the only way to obtain a good look is to play a recording of its call back to a singing bird. Being highly territorial, the male scrub-bird creeps close then runs rapidly around the tape recorder, in and out of the vegetation, stopping only long enough to sing a few quick notes.

The female is even more of a problem to locate. She does not sing and appears to spend most of her time in a different part of the territory than her mate. Though the Rufous Scrub-bird is sedentary, identifying its territory does not guarantee finding the bird; and the difficulty of locating a nest is almost legendary. Those that have been found have been near the edge of a male's territory, and tended

Distribution
Rufous Scrub-bird.

Robert Edden 1983.

primarily by the female. The nest is a globular structure of dry leaves with an entrance at the side. Its interior is lined with a wet layer of wood pulp and decayed wood, which dries to a cardboard-like consistency. Two eggs, pale pink with reddish spots, are cared for by the female. Because of the difficulties of keeping this bird under observation, its breeding biology is not well known.

Rufous Scrub-birds eat small invertebrates, supplemented by seeds, found among the debris collected on the forest floor. While foraging, these birds may disappear into the fallen litter and move short distances under the surface like a small rodent.

The lyrebird has long been considered the nearest relative to the scrub-birds. Although quite different in many outward respects, these two groups of birds do share some behavioural and anatomical similarities. They are the only birds which have a particular arrangement of the muscles of the voicebox or syrinx. Scrub-birds are unique among songbirds in that their collarbones do not join to form a wishbone.

The only other species of scrub-bird is the Noisy Scrub-bird *A. clamosus*, long thought to be extinct until rediscovered in 1961 in very small areas of Western Australia. At one time the distributions of both species were undoubtedly more extensive.

Opposite: Rufous Scrub-bird *Atrichornis rufescens*.

Robert Edden 1983

Chapter 15

CUCKOO-SHRIKES

(Campephagidae; Passeriformes)

Cuckoo-shrikes and the closely related trillers form a distinctive family of uncertain relationships. Australia has eight species, three of which occur in rainforest on a regular basis. With their slender grey bodies and buoyant flight, they superficially resemble the unrelated cuckoos. The hooked beaks are somewhat similar to those of true shrikes which do not occur in Australia. Elsewhere the name 'greybird' is often used for these birds.

Barred Cuckoo-shrike

The Barred Cuckoo-shrike *Coracina lineata* receives its name from its boldly marked underside; it is also called the Yellow-eyed Cuckoo-shrike for obvious reasons. The bulk of the diet — about 90 percent — consists of fruit and the balance of insects. It is an active bird, joining other fruit eating species in trees where food is available. It selects fruit of a size suitable to swallow whole. A partiality for wild figs has been shown, and a flock will eat as many of these fruits as it can before the fruit is infested with a small beetle. Once the fruit is attacked by the beetle the birds will not eat it. Barred Cuckoo-shrikes also congregate into single species flocks of up to 50 birds which travel through the forest, quickly moving from tree to tree, foraging on available resources. As they wander, a chattering 'aw-loo-ack aw-loo-ack' is given, often in flight. Many individuals gather at the same roost at night.

The range of this species extends from northern New South Wales to Cape York with other sub-species occurring in New Guinea and the Solomon Islands. It resides in rainforest and neighbouring eucalypt forest, at times visiting more open habitats. This species appears to be nomadic within its range but its movements require further study to be properly understood. Nowhere does this bird seem to be common.

Cuckoo-shrikes build noticeably small nests for their size. A small flat cup, placed on a high horizontal branch, is fashioned from small twigs bound together with spiderweb, sometimes decorated on the outside with lichen.

Distribution
Barred
Cuckoo-shrike.

Cicadabird

Less conspicuous than the Barred Cuckoo-shrike is the sombre-plumaged Cicadabird *C. tenuirostris*. For much of the time, it remains in the upper foliage of the forest. Unlike its relative, this species does not form prominent flocks. Instead single birds or pairs move quietly through the leaves, about a metre inside the canopy, in search of insects and fruit. Nor is the Cicadabird as noisy as the Barred Cuckoo-shrike. It produces an unusual buzzing trill similar to the call of a cicada. A colourful nickname of this species is Jardine Caterpillar-eater, now rarely used.

Distribution
Cicadabird.

Opposite: Barred Cuckoo-shrike *Coracina lineata*.

Whereas in most of Australia's cuckoo-shrikes the sexes are alike in appearance, male and female Cicadabirds are quite different. He is slate-grey and black while she is dark brown above, paler with dark barring below; her wing feathers are prominently edged with buff or cream. Breeding begins in August-November when adults arrive from the north. In February-April they start their return migration. The extent of these northward movements are not yet known.

The nest is similar to that of the Barred Cuckoo-shrike. Both parents construct the nest 8 metres to 30 metres above the ground. Only the female, fed by her mate as she sits, incubates the single egg.

Cicadabirds also range from Sulawesi through New Guinea to the Solomon Islands, showing considerable variation in the colour of the female plumage between regions.

Varied Triller *Lalage leucomela* — male (left); female (right).

Distribution Varied Triller.

Varied Triller

Trillers are miniature cuckoo-shrikes. Their name refers to the harsh, churring trill which is their typical call. The Varied Triller *Lalage leucomela*, common through its Australian range, is also found in New Guinea.

For its size, this species has the smallest nest and largest egg of any Australian member of its family. The dainty structure, usually placed in a small fork, is 70mm in diameter, 30mm high and only 10mm deep in the bowl, where the female deposits a single spotted egg. Nesting activity peaks from September to December with pairs holding small territories at this time. Throughout the remainder of the year they make local movements but are generally sedentary.

Varied Trillers join other species in mixed flocks, moving through the forest in search of food. Fruit, seeds, insects and spiders are taken. Trillers are not shy birds but may be difficult to observe in the rainforest canopy. They can be located by their distinctive trill when they move into less dense habitats.

Female Varied Trillers resemble female Cicadabirds but are smaller and have prominent wing bars and eyebrows. Males are boldly patterned with glossy black and white.

Chapter 16

GROUND THRUSH

(Turdidae; Passeriformes)

The Ground Thrush *Zoothera dauma* is spread from Australia to southeast Asia and India, north to Siberia and occasionally eastern Europe. Through this extensive distribution, it has acquired a variety of names: Scaly Thrush, White's Thrush, Tiger Thrush, Golden Mountain Thrush, Speckled Thrush, several of which are of use in Australia. The scalloped pattern of half-moons accounts for some of these titles. These markings help the Ground Thrush blend with its surroundings. This, and its quiet demeanour, make it inconspicuous on the shaded forest floor.

If disturbed, the thrush runs a short way or flies to a low branch and freezes. Generally its coloration conceals it from the observer. The first glimpse of this shy species is often the flash of the boldly marked black and white underwing, visible only when the bird flies. With patience and care, it is possible to watch Ground Thrushes as they feed. One or two birds move quietly among the leaf litter, picking up insects, worms, small snails and fruit with jabs of their bills.

The Ground Thrush starts breeding in winter and continues into the summer. It places its nest up to 15 metres from the ground, on a stump, in a fork of a tree or between the trunk and some loose bark. A rounded bowl of small roots and strips of bark is covered on the outside with moss and lined on the inside with fine pieces of vegetation. It holds two to three pale green or off-white eggs with variable amounts of red-brown markings.

Distribution Ground Thrush.

In Australia, the Ground Thrush resides in rainforest and wet eucalypt forest, moving to more open country when not breeding. Recent studies have suggested that the Ground Thrush in this country is actually composed of two very similar but distinct species. Northern birds have a richer buff-bronze colour and quite different songs than southern birds.

ROBINS

(Pachycephalidae; Passeriformes)

The introduction of the term 'robin' for some of Australia's insectivorous birds comes from their outward similarity to the unrelated but well known species of Britain and Europe. Like the European Robin *Erithacus rubecula,* they are plump, large-headed birds which feed on insects, usually by pouncing on them from a lookout point. Many have coloured (often red) breasts. Robins from both parts of the world perform the same roles in the environment which explains why they have come to resemble one another. The robins of the Australian region are more closely related to other insect-eating birds of this part of the world. They display a range of brightly coloured underparts as well as softer hues. Inquisitive and confiding, they allow close approach, frequently moving to within a few metres of a person in order to watch them. Whether observing humans or searching for insects, robins commonly adopt a characteristic pose, hanging sideways on the trunk of a tree or on a vine. At this vantage point, they remain motionless for long periods before darting down to capture prey or flitting off to another perch.

Eastern Yellow Robin

There is at least one species of robin in every part of Australia, except in the most arid sections, and some use a range of habitat types. Among the most versatile is the Eastern Yellow Robin *Eopsaltria australis.* It is one of the commonly seen rainforest birds, yet is also at home in the dry mallee, in parks and gardens and in open woodland. Eastern Yellow Robins of the coast and highlands from northern New South Wales through Queensland have a bright yellow rump. This makes a flash of colour as the bird darts through the gloom of the rainforest. Southern birds and those in drier areas have duller olive rumps

which contrast less with the back and tail. Adults of both sexes are alike but individuals having just left the nest are rufous with pale streaks.

Eastern Yellow Robins start to nest early in the season, June to July, and continue through to January or February. A pair may raise up to three broods within this extended period. Mated robins remain in the same area for several years. Two to three blue-green eggs with red spots are laid in a cup-shaped nest built in the fork of a tree. The chicks hatch 15 days later and remain with the parents for several months, forming small family groups that are known to assist with the rearing of subsequent broods. There is no evidence of a pronounced seasonal movement northwards after the cessation of breeding but in winter, Eastern Yellow Robins shift from higher elevations into the lowlands.

Yellow Robin at nest.

Opposite: Ground Thrush *Zoothera dauma.*

This species, nicknamed Yellow Bob, is a favourite and familiar bird of forests and gardens. In many areas it will frequently visit picnic tables when people are around and it has learned to accept handouts of cheese. The Eastern Yellow Robin is one of the first birds to become active in the morning and one of the last to go to bed at night. The name *'Eopsaltria'* means 'dawn harper'. The song is a series of monotonous but not unpleasant piping notes. Like other robins, this one has a harsh scolding call given when disturbed or annoyed.

Distribution Eastern Yellow Robin.

Pale Yellow and White-faced Robins

Overlapping the Eastern Yellow Robin in much of its rainforest distribution, the Pale Yellow Robin *Tregellasia capito*, at first glance, resembles it closely. The yellow legs, dull rump and pale face of the latter help tell them apart. In general behaviour, the two species are similar except that the Pale Yellow Robin tends to be a more active feeder and somewhat more timid. When foraging, it may pick insects off leaves or dart into the air, but most often it searches and pounces. This species is more restricted to rainforest and heavily wooded situations, avoiding open habitats. The voice is an undistinguished collection of short squeaks, soft trills and scolding chatters.

There is a large gap in the distribution of the Pale Yellow Robin. It is uncertain why this species is not present in seemingly suitable habitats in mid-eastern Queensland. Northern birds are noticeably smaller than those to the south and, from reports, appear to be more common.

It has been pointed out that there is a strong correlation between the occurrence of the Pale Yellow Robin and the presence of the rainforest plant, Lawyer Vine. These birds are extremely reliant on this vine, using it in the construction of their nests, both for material and for a site. Grass, small roots and fibre from Lawyer Vine are woven into a small cup, bound with spiderweb. It is usually placed in the junction of the stem and a leaf of this vine within 2.5 metres of the ground. The nest lining, too, is made from Lawyer Vine fibre. Juveniles are brick red with a few pale streaks on the crown. Pale Yellow Robins are sedentary.

In New Guinea and far northern Cape York Peninsula, it is replaced by the White-faced Robin *T. leucops*. This bird is closely related, the most obvious difference being the sharply marked facial pattern. The large white area broken only by black at the top of the bill has led one author to remark that it resembles a giant panda's face. The White-faced Robin is noted for its tameness but otherwise is very much like the Pale Yellow Robin in its natural history.

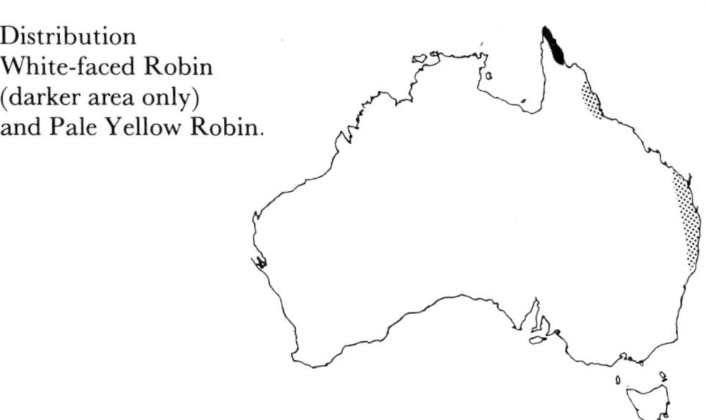

Distribution White-faced Robin (darker area only) and Pale Yellow Robin.

Grey-headed Robin

The largest of Australia's robins is the attractive Grey-headed or Ashy-fronted Robin *Heteromyias cinereifrons*. It has a tortoise-shell pattern to its plumage that is unlike any other robin except for a close relative in highland New Guinea. The Grey-headed Robin is also a highland bird and within its narrow range is found almost exclusively above 400 metres. Here it is a common, sedentary species. It frequents the edges of clearings, tracks and other openings in the rainforest more than the closed interior. Roadsides are among the best places to view this bird as it darts out from the forest to pounce on an insect before hurrying back to cover. The behaviour of the Grey-headed Robin is like that of other robins but it also hops along the ground on its proportionally longer legs. One of the characteristic sounds of these highland rainforests is the whistled song of this species, one long note followed by three lower ones. In quality, the song is reminiscent of the piping of the Eastern Yellow Robin. The nest is an untidy cup set one to two metres above the ground.

Opposite: Eastern Yellow Robin *Eopsaltria australis*.

Robert Edden 1983

Rose and Pink Robins

Three of the robins discussed thus far are yellow-breasted forms. Another prominent group includes those with red or orange breasts. These behave in much the same manner but one, the Rose Robin *Petroica rosea,* is considerably more arboreal in its habits. It feeds high in the canopy, actively pursuing insects in the outer foliage. Frequently the Rose Robin dashes into the air to capture prey on the wing and the manner in which it fans its tail as it does brings to mind the fantail flycatchers. This species is the smallest of Australia's robins; it is also slimmer in body and, for its size, longer in the tail, giving it less of a robin-look than its relatives.

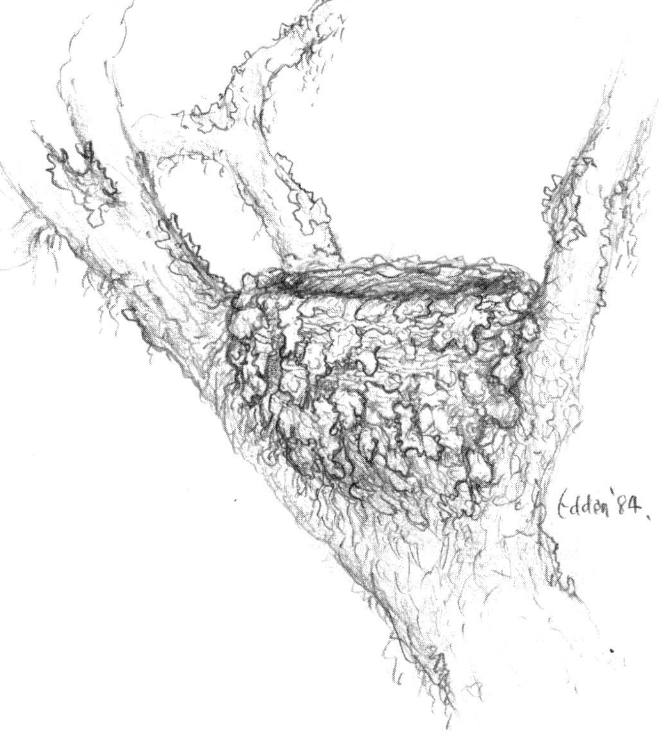
Rose Robin nest.

The male Rose Robin is unmistakable but problems arise separating the brown robin female of this species from closely related forms. Immature males have a female-like plumage but are able to breed before acquiring the attractive adult colours. During the breeding season Rose Robins remain in densely vegetated gullies, wet eucalypt forest and rainforest. Males have a trilling 'dint dint dint deer deer' to advertise their territories, and a 'tick tick tick', like the snapping of small twigs. Females build a small, deep cup of moss and fibre, taking about two weeks to complete the job. This is placed in a

fork or on a horizontal branch, up to 20 metres from the ground. The outside of the nest is heavily covered with lichens. Rose Robins may raise up to three sets of young in a season, laying two to three eggs each time. During the two week incubation period, the male feeds his mate as she sits on the nest. They later join to bring food to the chicks. Rose Robins may be partly migratory; certainly there is a post-breeding dispersal into the more open lowlands.

A peculiar behavioural trait of the Rose Robin and its close relatives is to droop and nervously flick the wings and tail when perched. Other robins do this occasionally but nowhere near as much as the red-breasted species.

Temperate rainforest gullies in Tasmania and southern Victoria support the Pink Robin *P. rodinogaster,* a close counterpart of the Rose Robin. Where they both occur, the Pink Robin can be distinguished by the black upperparts of the male and the absence of white in the tail of either sex. Sporadically this species is recorded as far north as Canberra and may even breed along the Victoria-New South Wales border. The factors influencing these irregular appearances and the movements of this species in general are not yet understood.

Distribution
Rose Robin.

Yellow-legged Flycatcher

Although related to the robins and superficially similar, the Yellow-legged Flycatcher *Microeca griseoceps* behaves quite differently. Discovered in Australia as recently as 1913, it was not until this species was observed feeding that its unrobin-like actions were noted. Up through the last few decades, it was even called the Little Yellow Robin. The Yellow-legged Flycatcher moves through the

---

Opposite: Grey-headed Robin *Heteromyias cinereifrons* (above) and Pale Yellow Robin *Tregellasia capito* (below).

Following pages: White-faced Robin *Tregellasia leucops* (p.90); and (p.91) Rose Robin *Petroica rosea* — female (above); male (below).

Robert Edden 1983.

foliage, frightening resting insects and then capturing them in flight. It remains in the canopy, never clinging to the sides of trunks or alighting on the ground. This arboreal, flycatching mode of feeding is reflected in the weak legs and flat triangular bill. In colour, the Yellow-legged Flycatcher resembles the Pale Yellow Robin but the ranges do not overlap, except for rare reports of the Yellow-legged Flycatcher from the Atherton Tableland. It also dwells in New Guinea.

But this is a poorly known species. Its nest was not discovered until 1969 (in New Guinea) and not described for Australia until 1979. The group of birds to which the Yellow-legged Flycatcher belongs is noted for their small nest, and this species is no exception. The nest is a small lichen-covered cup, 4cm wide and 4cm deep, placed on a horizontal limb.

Yellow-legged Flycatcher *Microeca griseoceps*

Distribution
Northern
Scrub-robin.

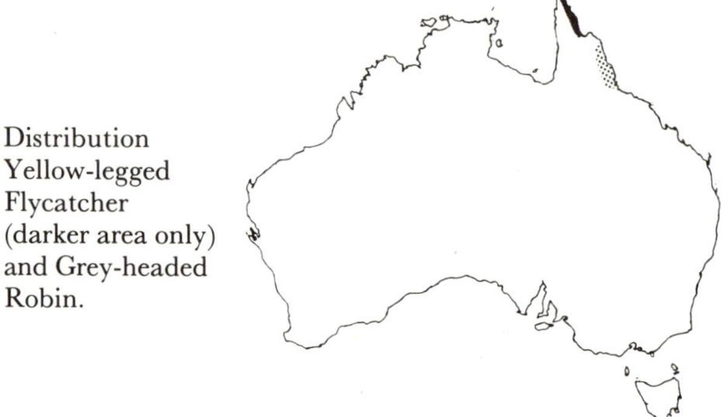
Distribution
Yellow-legged
Flycatcher
(darker area only)
and Grey-headed
Robin.

Northern Scrub-robin

The Ground Thrush is the only native species of true thrush found in Australia. For many years the scrub-robins were also considered thrushes but recent evidence indicates that they are not closely related. It is now believed that scrub-robins are large relatives of the more arboreal native robins but which have taken to an increased terrestrial existence. The mistake is understandable for they are very thrush-like in appearance and habits. One species of scrub-robin lives in the dry mallee country. The other — the Northern Scrub-robin *Drymodes superciliaris* — is restricted in Australia to a small part of northern Cape York Peninsula and around the Roper River in the Northern Territory. The Roper River form is known only from three specimens and has not been recorded for many years. It may be extinct.

Considered one of Australia's least known birds, the Northern Scrub-robin is no better known in New Guinea where it is more widespread. What has been observed of its behaviour is similar to that of its more familiar mallee relative. Both move across the ground with a hopping, bounding run and raise and lower the tail when perched. The call, a drawn-out whistle, is given from a low branch. Insects are the main food.

The Northern Scrub-robin of Cape York is sedentary. It starts to breed in October, finishing in January. A depression made at the foot of a tree or shrub and lined with sticks and smaller vegetable matter serves as the nest. Two light grey eggs with a thick covering of brown blotches are deposited.

Northern Scrub-robin *Drymodes superciliaris*.

Chapter 18

WHISTLERS AND SHRIKE-THRUSHES

(Pachycephalidae; Passeriformes)

The name 'whistler' is an accurate description for this group of birds, well known for their melodious, ringing songs. The songs differ from species to species but their richness of quality always identifies these birds. There are 28 species of whistler, occurring throughout the Australasian region and south-west Pacific. A few occur in the Philippines and as far west as India. All are similar in general behaviour and in appearance. Whistlers search for insects in a more leisurely and deliberate manner than most related insectivorous species. Translated, the generic name *Pachycephala* means 'thick head', a title that was used for these birds prior to the adoption of the more complimentary 'whistler'.

Golden Whistler

The most widespread and best-known species is the Golden Whistler *Pachycephala pectoralis*. The showy male is one of the most beautifully coloured birds in the eastern and southern mainland and Tasmania. This species is renowned for having more recognised geographic variation than any other bird. Between Indonesia and Fiji, the Golden Whistler has evolved around 80 different plumage forms. It has evolved 14 alone in its narrow Australian range. From area to area, the plumages of male and female have undergone independent changes. Males may have white throats or yellow throats, green backs or black backs; the yellow collar and black breast band may be absent or present. On some islands males exhibit a female or hen-plumage, lacking any of their typical bold pattern. Elsewhere females more closely approach males in appearance.

Variation in Australian birds is not as dramatic. In males the tail varies from grey to black, while Tasmanian birds have shorter bills than those of the mainland. Among east coast populations found in rainforests, it is the female which displays geographical differences. The male remains the same. South of the Hunter River, New South Wales,

females are grey with little suffusion of other colours in the plumage. Once past the Hunter River valley, an olive tinge appears in the upperparts, the underparts take on a buff colour, and a lemon yellow pervades the undertail coverts. Northwards the major changes from this pattern are in the greens, olives and browns of the back.

Golden Whistlers are primarily birds of the more humid forests, including rainforests, but extend inland to drier acacia scrubs, such as brigalow and mallee. This species is often seen or heard in

Golden Whistler at the nest.

93

parks and gardens in towns. Both sexes sing throughout the year, a loud disturbance like a thunder clap or car backfiring often setting them into song. There are several long phrases, difficult to express in words. One of the most frequently heard can be rendered 'che-che-che-too-wit' with a rising emphasis like a whip crack on the final note.

The breeding season brings out the best of these songsters as a pair of Golden Whistlers give the impression of trying to outdo their neighbour in volume and quantity of song. A display of see-sawing postures initiates courtship, after which an open cup of rootlets, grass and stems is built in a forked branch within a few metres of the ground. Fledglings are covered with brick red feathers, which is soon replaced by the female-like immature plumage — except for feathers of the wing which retain reddish edges. In the next moult, these too are lost and young males become almost indistinguishable from adult females. It is at least three years before males acquire the handsome black and gold plumage. Hen-plumaged immatures sing almost as much as adults and frequently breed before attaining the adult colours. At the end of the breeding period there is a degree of dispersal; some birds move out of the higher country and younger birds often winter to the north of their breeding range.

Distribution Golden Whistler.

Olive Whistler

More limited in distribution and habitat than the Golden Whistler — and less familiar to most people — is the Olive Whistler *P. olivacea*. It has a patchy distribution along the southeast and in Tasmania. The Tasmanian birds occur in open timbered habitats as well as cool temperate rainforest, including those at lower altitudes. On the mainland, this species is more restricted in its choice of habitats, reaching its greatest abundance between 1000 metres and 1500 metres. It is increasingly uncommon the further north one looks. The Olive Whistler prefers dense, humid forest, and has a strong as-

sociation with stands of Antarctic beech. Dense regrowth in partially logged forest at high altitudes is a favoured habitat in northern New South Wales. There is a post-breeding dispersion to more open country.

Olive Whistler *Pachycephala olivacea.*

Though larger than the Golden Whistler, the Olive Whistler lacks its striking sexual dimorphism. But as a songster it rivals its relative and, to some people's minds, exceeds it. The voice of the Olive Whistler varies geographically. The two major variations are a 'tu-we-tchow' in the south and 'peer-ah peer-ah peer-ah' in the north. Northern birds also have a distinctive 'peee-poo' call that is absent in southern populations, and in which the last note is much lower than the first. In other parts of the range some of these songs seem to resurface again, making the pattern of variation more complex than once thought.

Singing birds often perch high in the forest though they do much of their feeding in the dense understorey and place their nests within a few metres of the ground. Olive Whistlers are more difficult to approach than Golden Whistlers, often moving ahead just out of sight of any observers attracted by their song.

Grey Whistler

A loud joyful song announces the presence of the

Opposite: Golden Whistler *Pachycephala pectoralis* — female (above); adult male (centre); immature/adult female (below).

Grey whistler *P. simplex* in the rainforests of north-east Queensland. Its most common song consists of two short notes followed by five or more whistles sometimes written 'catch a fish alive'. Its habit of spending most of its time feeding in the canopy makes it hard to detect, as does its nondescript coloration. There is considerable variation in the amount of yellow on the feathers of the underparts and this may cause confusion with other species such as female Golden Whistlers and some robins or flycatchers. Sexes are similiar. Grey Whistlers actively hunt for insects among the leaves, now and then fluttering up to snatch one off the surface of the foliage. Other subspecies occur in the Northern Territory, New Guinea and the Moluccas.

Distribution
Grey Whistler
(darker area only)
and more southern
Olive Whistler.

Rufous Shrike-thrush

Shrike-thrushes are like larger versions of whistlers but without the bright colours found in some of those species. One group grades into the other and diagnostic characters are not clear cut. Like whistlers, shrike-thrushes have rich, beautiful voices. Their song phrases are shorter than those of whistlers but more mellow and resonant. They are predatory birds which eat a selection of small animals. The hooked shrike-like beak and thrush-like build gives this group its name; it is related to neither of these principally non-Australian groups. The misconception is perpetuated by the practice of abbreviating the name to 'thrush'.

The smallest species of this genus is the Rufous Shrike-thrush *Colluricincla megarhyncha*. Its size and many colour variations throughout its range (some are not rufous), has led to the introduction of the replacement name, Little Shrike-thrush. It is found outside Australia in New Guinea, Sulawesi and many of the surrounding islands. The Rufous Shrike-thrush has a discontinuous distribution from the north east across the top end and down the east coast to northern New South Wales. Breaks occur where the closed forest types (rainforest and man-groves) are interrupted by drier, open country. Birds of the north west are browner, have larger bills and occupy mangroves and monsoon forest; rufous

Rufous Shrike-thrush nest.

coloured eastern birds are found in rainforest. Between these forms are transitional populations, intermediate in characteristics.

First impressions can mislead an observer into believing that the Rufous Shrike-thrush is rarer than it really is. It searches unobtrusively for food on trunks and branches, in masses of vines or on the ground, either singly, in pairs or with large mixed-species flocks. The observer may well find himself being quietly watched by a Rufous Shrike-thrush from a nearby perch. When in song, this species is much more obvious. It has a range of harsh wheezy notes and clear piping whistles; the most commonly

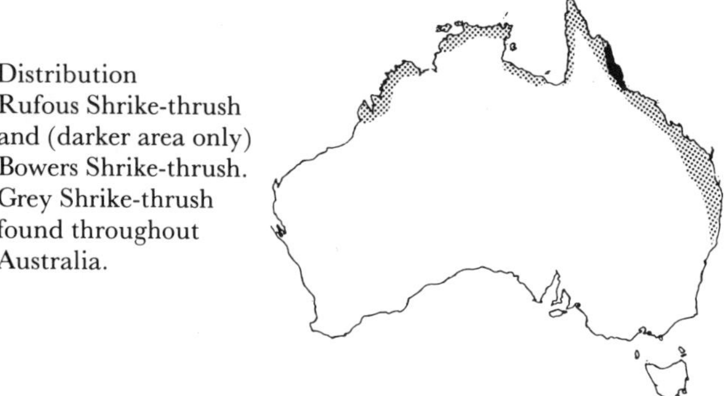

Distribution
Rufous Shrike-thrush
and (darker area only)
Bowers Shrike-thrush.
Grey Shrike-thrush
found throughout
Australia.

Opposite: Rufous Shrike-thrush *Colluricincla megarhyncha*.

Robert Fdden 1983.

heard phrase sounding something like 'a cup of tea, wot wot wot'. Rufous Shrike-thrushes breed September through February, and lay two to three eggs.

Bower's Shrike-thrush

The most restricted species, both geographically and altitudinally, is Bower's Shrike-thrush *C. boweri*. Limited to the Atherton Tableland district, this bird is resident in the highlands above 400 metres. Separation from the Rufous Shrike-thrush, which also occurs in these regions, is possible by the grey back and streaked front (hence the alternative name, Stripe-breasted Shrike-thrush) and heavier, darker bill. Its song consists of four quick notes followed by three ringing ones: 'da dee da dee pon pon pon'. Bower's Shrike-thrush is generally more silent than its relatives. It quietly forages in the lower layers of the forest, sometimes moving on to the lawns of adjacent houses in search of prey.

Within its range, Bower's Shrike-thrush is common. It begins to breed in October, situating its nest — a cup of leaves and bark — in a small tree fork or in a tangle of vines, three to eight metres above the ground. Its two eggs are white with small reddish-brown spots, particularly near the larger end.

Grey Shrike-thrush

Unlike the previous species with its limited range, the Grey Shrike-thrush *C. harmonica* is one of Australia's most widely distributed songbirds, residing in a wide variety of timbered habitats from rainforest to the scrubs of the dry inland. It has several distinct regional forms but only the eastern birds are found in rainforest. In eucalypt forest and woodland, it is perhaps more common and it frequently visits gardens, parks and homesteads. In the east it has grown tolerant of humans, nesting near houses and even entering buildings if accustomed to receiving handouts. An inquisitive species, the Grey Shrike-thrush is often attracted to a vantage point from which to watch the activities of humans.

In the Grey Shrike-thrush, this group of songsters reaches its pinnacle. The powerful, melodious songs of this species have made it a familiar and well-liked bird throughout Australia. There are numerous regional dialects and considerable in-dividual variation in its many songs, perhaps as wide a range as any native songbird. Some of these

Grey Shrike-thrush *Colluricincla harmonica*.

have been expressed as 'yo-ho-hee', 'pip-pip-ho-ee' and 'pur-pur-purquee-yule'. The nicknames of the Grey Shrike-thrush reflect the impression its voice has made: Harmonious Thrush, Whistling Dick, and in imitation of some of its calls, Joe Wickie. One part of its repertoire has been interpreted as 'Duke-Duke-Wellington' and inevitably spawned yet another nickname. Its winter call is a loud, downward 'chong'.

The Grey Shrike-thrush acts more like true thrushes than the rest of its genus. More terrestrial than the others, it hops across the ground seeking insects or searches along larger branches, tree trunks and fallen logs. Frogs, small lizards and mammals, eggs and nestling birds round out its diet.

The nest is an untidy bowl of coarse material such as strips of bark, thick grass and other plant fibre. It may be placed on the ground, in a tree fork or in a hollow limb or stump. Nests are even built in a niche in a building. Two to four, usually three, spotted eggs are laid. Adult males have a black eye ring, white spot in front of the eye and a black bill; females have a white eye ring, grey bill and grey spot in front of the eye. Grey Shrike-thrushes are sedentary other than some local or altitudinal movements in the non-breeding season.

Chapter 19

MONARCH FLYCATCHERS

(Pachycephalidae; Passeriformes)

The monarchs are a group of colourful, attractively marked flycatchers that are extensively distributed from Indonesia east through the islands of the South Pacific. They do little 'flycatching' (capturing flying insects while in the air), however, preferring to secure their prey from the surfaces of leaves, branches and trunks of trees. Several species may occur together but separate from each other by the section of the forest in which they choose to forage. All have harsh, scratchy call notes and distinctive, though not musical, songs. A characteristic of many of the monarchs is their slate blue-grey bill and legs, a colour that is rare among songbirds of the world.

Black-faced and Black-winged Monarchs

The loud whistle of the Black-faced Monarch *Monarcha melanopsis* is a common sound of the rainforests, dense eucalypt forests and wet gullies during spring and summer. It can be written as 'why-you-which-yeew', with the long third note sharp and rising. The bird moves sedately through the middle layers of the canopy, usually in the interior of the trees, in search of insects.

In October, Black-faced Monarchs begin to pair up in preparation for breeding . Their nest is a well made cup of rough plant fibre, often including the needles of the She-oak, liberally covered with green moss. The female lays and incubates two or three red-spotted eggs. Adults appear alike but immatures lack the black face. At the close of the breeding season around March, these monarchs begin to migrate north. Southeastern populations seem to vacate their area entirely, while some birds are present thoughout the year in the north. In southern New Guinea, they appear in winter, indicating that there is a migration across Torres Strait. They return in August and September. The Black-faced Monarch is most common in the northern parts of its distribution, around Cooktown, and decreases in abundance the further south one goes. Rarely is it found west of the Dividing Range.

A very similar species, which exists on the tip of Cape York Peninsula is the Black-winged Monarch *M. frater.* For many years it was debated whether this species was the same as the Black-faced Monarch and whether it remained in Australia throughout the year. The Black-winged, or Pearly Monarch differs from the preceding species primarily by having black wings and tail instead of grey. Calls, breeding and feeding behaviour are alike for both. Although the Black-winged is not well studied, it has been found that where these two species exist together in New Guinea, they occupy different altitudinal zones with little or no overlap. The Black-winged Monarch prefers the higher altitudes. Also where the Black-winged Monarch is found on Cape York Peninsula, the Black-faced Monarch seems to appear only while on migration and is not known to breed there. For these reasons, ornithologists consider the Black-winged Monarch to be a distinct species, rather than a distinctive colour form of the Black-faced Monarch. Part of the Cape York population migrates back and forth to New Guinea each year.

Distribution Black-faced Monarch.

Spectacled Monarch

The beautiful Spectacled Monarch *M. trivirgata* is similar in colour to the Black-winged and Black-faced Monarchs but it is slimmer and more active as it seeks its food in the outer branches of the lower canopy. Fluttering through the foliage, it behaves somewhat like a fantail, though it does not chase flying insects on the wing. By fanning its tail and darting about the leaves, this monarch flushes resting insects. As it moves, the prominent white tips of its tail feathers are displayed. In addition to harsh chattering notes, the Spectacled Monarch makes a buzzing 'swee swee swee', often repeated several times.

Like many species which are spread along the east coast, the Spectacled Monarch exhibits different types of annual movements between the

whiter on the underparts and this subspecies has been called the White-bellied Flycatcher.

The nest, like those of the other monarchs, is a sturdy cup placed in an upright fork. Its shape varies according to its situation as it is built to fit tightly between the twigs which support it. Two eggs are laid during the breeding season which lasts from October to January.

Distribution Spectacled Monarch.

Distribution White-eared Monarch and (darker area only) Black-winged Monarch.

White-eared Monarch

Often sharing the same tree as Black-faced and Spectacled Monarchs, the White-eared Monarch *M. leucotis* avoids competition with them by actively moving across the surface of the canopy after prey. Unlike its larger relatives it also springs into the air to capture passing insects.

Seasonal movements of the White-eared Monarch are not fully understood. It may be migratory in the southern part of its range and sedentary elsewhere. Part of the problem of studying this bird is that it is an inhabitant of the upper layers of the forest. Even its nest, placed high in a tree, was not identified until 1923, 70 years after the species was first discovered. In the right areas, the White-eared Monarch can be observed without great difficulty, but it does not seem to be as abundant as other

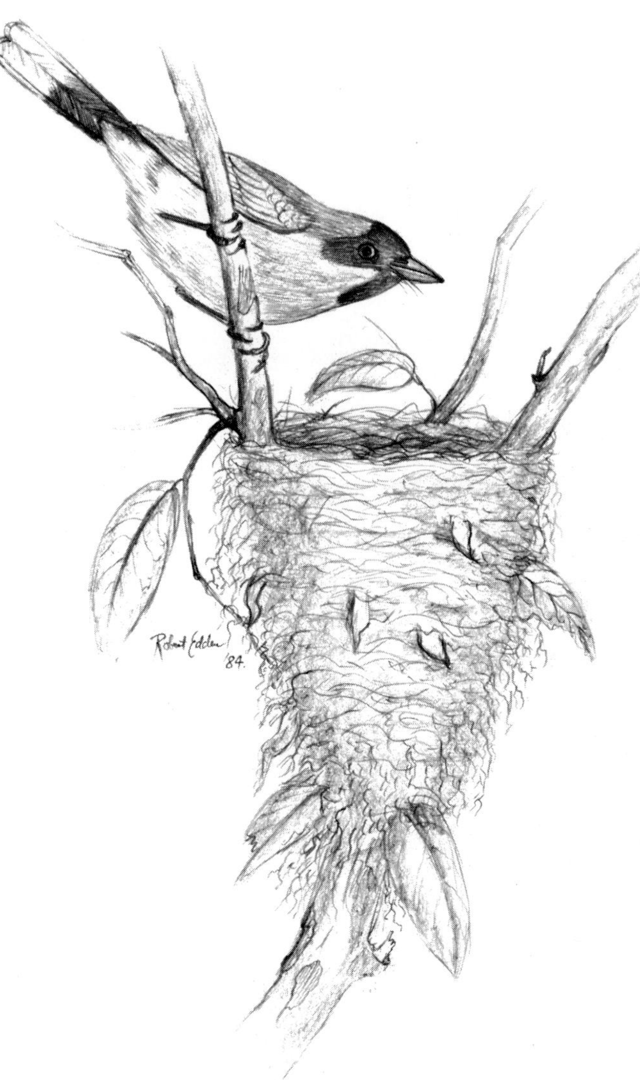

Spectacled Monarch at the nest.

north and south. As in the Black-faced Monarch, birds which breed in the south migrate north, while northern birds are sedentary. On Cape York Peninsula, the birds of the breeding population are much

Opposite: Spectacled Monarch *Monarcha trivirgata* (above); and the Black-faced Monarch *Monarcha melanopsis* (below).

White-eared Monarch *Monarcha leucotis*.

monarchs. Unlike them, it is confined to Australia; nearly identical forms, however, are found on some small islands of eastern Indonesia. This species' most common calls have been described as a drawn out 'eee-choo eee-choo' and a whistling 'you get away', accented on the second note. Young birds are grey and white and have a dull rufous wash across the breast.

Pied and Frilled Monarchs

Two closely related but very distinctive species are the pied Monarch *Arses kaupi* of the Atherton district and the Frilled Monarch *A. telescophthalmus* of northern Cape York Peninsula, New Guinea and satellite islands. These are separated from the typical monarchs by their erectile frills, coloured eye wattles, frail hanging nests and mode of feeding. In Australia there is little sexual dimorphism and the most obvious difference between the two species is the black breast band of the Pied Monarch. As this band starts to disappear in the northern part of its range where it approaches the Frilled Monarch, the suggestion has been made that perhaps they are the same species. New Guinea populations of the Frilled Monarch show striking differences between the sexes — females are bright rusty red throughout much of their plumage. Both species are sedentary within their restricted Australian ranges. The Frilled

Flycatcher seems less confined to rainforest and is often seen in adjacent eucalypt forest.

The erectile feathers on the nape can be raised into a frill. This is more obvious in the Frilled Monarch and is often seen as the bird moves through vegetation and during courtship displays. At times small parties call and chase one another with frills erect in a group display of unknown significance.

The delicate nest consists of a small cup of thin twigs and vine tendrils. It is suspended between two vines like a hanging basket, some six to ten metres from the ground. Two eggs, speckled like those of the other monarchs, make up the clutch.

Recent studies of the Frilled Flycatcher in New Guinea have demonstrated that the differences in feeding behaviour between the male and female are as great as those that might be expected between quite separate species. The male spends most of his time at the height of the understorey, searching along bare surfaces like tree trunks, large branches or lianes. He moves about the bark, spiralling like a treecreeper. His mate forages in the lower canopy where the vegetation is more open. Here she chases flying insects or hunts among the leaves for prey. Related to these distinct methods are differences between the sexes in morphology. To assist with climbing about the bark, the claws of the male are larger than the female's. She, on the other hand, has a longer tail, which gives her greater maneuverability in her aerial pursuits. She also has larger bristles around her bill, which helps her to capture insect prey.

Distribution Frilled Monarch (darker area only) and Pied Monarch.

Shining Flycatcher

Described by some as Australia's most attractive flycatcher, the Shining Flycatcher *Myiagra alecto*

Opposite: Frilled Monarch *Arses telescophthalmus* (above); and Pied Monarch *Arses kaupi* (below), both adult males.

Boat-billed Flycatcher *Machaerirhynchus flaviventer* — male (left); female (right).

receives its name from the glossy black male. The female is equally striking, though in a different way — a shining black crown and white throat and underparts, contrasting with a rufous back and tail.

More a bird of the mangroves, the Shining Flycatcher rarely strays far from water and may also be found in paperbark swamps and in gallery rainforest along rivers and streams. It feeds more actively than the previous species, sometimes springing from an exposed perch to capture a flying insect, but more often foraging low in the forest. In the low undergrowth or on the ground, this flycatcher darts after moving prey or picks up small crabs and shellfish. When perched, it regularly quivers its tail or swings it side to side. The Shining Flycatcher has several calls, one like the croak of a frog, another like 'a man whistling a dog'.

Distribution Shining Flycatcher.

Boat-billed Flycatcher

Perhaps most closely related to the preceding group of flycatchers is the Boat-billed Flycatcher *Machaerirhynchus flaviventer*. It is an extraordinary bird. Its

alternative names, Yellow-breasted Boatbill or Flatbill also refer to its bizarre flattened bill, which is described as 'canoe-shaped', hence its name. However, despite its size, this bill is very weak and can only cope with the softest bodied insects. Boat-billed Flycatchers capture prey off leaves or in the air. They forage actively in the lower and middle layers of the forest but it can be difficult to spy them amid the foliage.

They are most conspicuous in the breeding season between September and March. Most often seen in pairs, the sexes differ slightly in that the female is duller than her brightly coloured mate. Their buzzing 'wit zee-ee-ee wit' is given frequently. The peak of vocal activity is reached by the male as he builds the nest, singing constantly as he carries out this task. Some observers use the level of his singing to determine the onset of nest building. The flat saucer of fine twigs is placed towards the end of a branch among the leaves anywhere from four to twenty metres high. Its construction, however, is so frail that the eggs can often be seen through the sides.

Two interesting behavioural traits of the Boat-billed Flycatcher have been noted. Unlike other flycatchers in Australia, this species carries its body

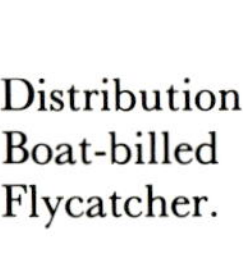

Distribution Boat-billed Flycatcher.

in a horizontal posture and when excited raises its tail in the air. Also small congregations of these birds have been seen participating in a group display, calling and flying back and forth along a path. No one understands the role of this ceremony.

The Boat-billed Flycatcher also occurs in New Guinea as does a related species with a black, instead of yellow, breast. They live at different elevations and do not come into contact. In Australia, the Boat-billed Flycatcher is found along the coasts and in the highlands and even on some nearby islands. It inhabits rainforest, occasionally venturing into adjacent woodland.

Chapter 20
FANTAILS
(Pachycephalidae; Passeriformes)

Grey Fantail

The familiar Grey Fantail *Rhipidura fuliginosa* must be one of the most popular of Australia's small birds. It quickly draws attention when feeding by its frenetic behaviour. With tail fanned, a bird sidles along a branch, putting small leaf-dwelling insects to flight. It will suddenly dash from its perch, pursuing them with acrobatic twists and turns, abruptly changing directions. Rarely does it fly in a direct path for more than a few seconds. These conspicuous chases have earned it such names as 'Mad Fan' and 'Cranky Fan'. Tame and inquisitive, Grey Fantails can be easily attracted by making a squeaky, kissing noise. They are fearless, attacking hawks and magpies that stray too close to the nest.

The Grey Fantail, represented by several distinct subspecies, is found throughout Australia, except for some of the most arid parts of the interior. Along the east coast and neighbouring highlands, this species occupies almost any habitat from rainforest, waterside vegetation and mangroves to gardens and open woodland. This adaptability doubtless contributes to its success. In the highlands of northeast and mid-east Queensland, lives a very dark, almost black, form which is mostly confined to rainforest. The Grey Fantail also inhabits New Zealand, New Caledonia, Vanuatu and the Solomon Islands. Rarely does it reach New Guinea where two very similar species replace it through most of the country.

Annual movements of the Grey Fantail need further study. It is probable that Tasmanian birds cross the Bass Strait in winter, however, the species' travels through the rest of the country are not well understood. Some migration is suspected but the picture is complicated by altitudinal movements after nesting. Southeastern birds shift from the ranges to lower elevations, often into dry, open country. Other post-breeding movements appear nomadic.

Regardless of what happens at these times, Grey Fantails reappear to breed in the same areas each year. By August most have arrived and nesting is under way. Fantails build a characteristic 'wineglass' shaped nest. A small, neat cup of thin strips of bark and fine grasses, bound with spiderweb, is fastened into a small fork of a branch. Hanging from

Grey Fantail *Rhipidura fuliginosa.*

Robert Edden 1983

the bottom is a tail like a baseless wine glass stem, some 150mm long. The structure, hardly large enough to accommodate the two to four eggs, seems increasingly inadequate as the chicks grow, the young birds at times threatening to spill over the edge. Grey Fantails are often parasitised by cuckoos which lay their eggs in the nest of the unsuspecting foster parents. The young cuckoo grows to several times the size of its fantail host.

The high ascending song of the Grey Fantail is a familiar melodious sound of the bush. It has been compared to a violin or fiddle.

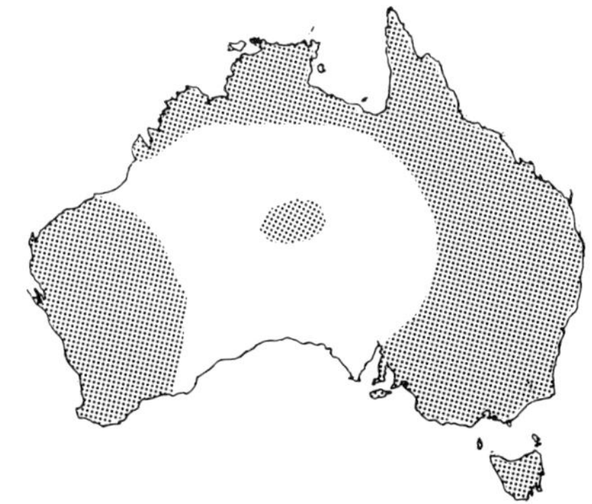

Distribution
Grey Fantail.

Rufous Fantail

The beautiful Rufous Fantail *R. rufifrons* resembles its grey relative in more ways than just in its general shape and habits, but each aspect has subtle individual differences. This species darts through the foliage picking insects off leaves or makes an aerial foray for a flying insect. It lacks many of the acrobatics of the Grey Fantail, capturing its prey with a more direct sally. Rufous Fantails generally feed lower in the vegetation, even chasing insects on the ground.

Widespread outside Australia (Sulawesi, New Guinea, the Solomon Islands, Guam), the Rufous Fantail is more restricted in its Australian distribution than the Grey Fantail. It is also more specific in its choice of dense vegetation and mangroves. However, during migration it appears in all sorts of open habitats, including gardens and city streets. This species has definite migratory movements though these vary throughout its range. In the southeast of the country all Rufous Fantails depart between February and April, though the young remain longer than the adults. Northern birds are partly migratory, with some apparently reaching New Guinea. There may be altitudinal movements as well.

In October, most Rufous Fantails have returned to their southern breeding grounds. As soon as they arrive, they begin to build a wine-glass nest similar to the Grey Fantail's though slightly larger and with a tail only half the length. When the young have departed, adults leave their territories. They move nomadically within their range until it is time to head northwards again.

Distribution
Rufous Fantail.

Opposite: Adult pair of Rufous Fantails *Rhipidura rufifrons*.

Chapter 21

LOGRUNNERS

(Orthonychidae; Passeriformes)

Southern Logrunner

The Southern Logrunner *Orthonyx temminckii* is a common resident of temperate and subtropical rainforest between Gympie, Queensland, and Illawarra, New South Wales. Encountering it, however, is not always a certainty. Sometimes its presence is not detected until the loud 'weet weet weet' cries are heard from nearby scrub in which it has been disturbed. On other occasions, this bird may be seen feeding on the forest floor when it is very tame, often almost oblivious to human presence. If startled, the logrunner moves a short distance away, but quickly returns to continue feeding, even if the observer remains in the same place. Foraging logrunners have been known to walk across people's shoes provided they remain still.

The feeding behaviour of this bird is interesting to watch. The large, powerful legs and feet are used to clear leaves and other debris from the ground, uncovering insects, crustaceans, snails and other small invertebrates which form the diet of the logrunner. Instead of the more usual backwards scratching motion of the leg used by many ground-feeding birds, logrunners have a strange sideways kick given with alternate legs. As they feed, they work in circles, leaving cleared patches about 150mm in diameter. The tail feathers have projecting centre shafts from which the name Spine-tailed Logrunner is earned. These sharp tips are pressed against the ground to support the bird as it rakes the debris aside.

Pairs or small parties, presumably family groups, are seen moving about the forest floor. Males and females differ by the colour of the throat and upper breast, females sporting chestnut-orange where males are white. Whether this contradicts the 'normal' case in which the male is the brighter of the pair in sexually dimorphic species is uncertain — the white throat of the male may seem conspicuous

among the shadows of the forest floor. Juvenile birds accompany the parents and can be recognised by their more mottled appearance. The throat region is also mottled but as the next feather covering emerges, the characteristic colour appropriate to the sex of the individual appears.

Logrunners are sedentary and maintain territories throughout the year. These they vigorously defend from neighbouring logrunners. Noisy territorial disputes involve adjacent pairs facing-off across a boundary: such arguments often provide the first clue to the location of this species. This aggressive defense allows them to be easily attracted by playing a recording of their call to which they quickly respond by returning the call and approaching the source. One territorial defender even attacked a tape recorder that was being played while carried on a person's back.

Within the territory, the female Southern Logrunner builds a large bulky dome nest about 220mm in diameter. Made of twigs and sticks, it is placed on or near the ground, often under overhanging moss or fern. Moss is placed around the entrance which opens to the side and has an approach consisting of a small platform of sticks. Two entirely white eggs are laid. The female alone incubates them and subsequently cares for the young in the nest. Unlike the majority of rainforest birds, this species is a winter nester, April to August being the normal breeding season.

The distribution of the Southern Logrunner is unusual. It is common throughout its habitat in the middle of eastern Australia; it is absent from north-eastern Queensland; and it reappears in highland

Southern Logrunner *Orthonyx temminckii* — female (above); male (below).

New Guinea where it is apparently rare and little known.

Northern Logrunner

Between these two disjunct localities — in the highlands of the Atherton rainforest tract — lives a closely related species, the Northern Logrunner *O. spaldingii*. It is half-as-big-again as its relative and

A male Northern Logrunner *Orthonyx spaldingii*.

most of its body is a solid black, not mottled brown. Plumage differences between male and female are the same, but these species show other distinctions in their breeding biology.. The Northern Logrunner may breed in any month of the year, though it tends to avoid the wettest periods and it lays only one rather than two eggs.

To call this bird 'northern' when the 'southern' species reappears to the north is a somewhat parochial approach but the names have been in use for some time. Perhaps other names, such as Black or Brown Logrunners, or Greater or Lesser Logrunners, would be preferable. Another title sometimes used for Northern Logrunner is Chowchilla. This refers to the loud call heard primarily at dawn and dusk. The echoing 'chow-chowchilla' is repeated, often drowning out many of the other forest noises. This species also mimics the voices of other birds.

One of the mysteries of the logrunners is which birds are their nearest relatives. Current

Distribution Northern Logrunner (darker area only) and Southern Logrunner.

classifications place them with the whipbirds and quail-thrushes, a diverse group of birds that seem to have little in common other than being terrestrial. The structure of the nest, division of parental duties and timing of the breeding season (at least in the Southern Logrunners) are remarkably like those of lyrebirds. Whether this interesting possibility is true awaits confirmation by further studies.

Opposite: Eastern Whipbird *Psophodes olivaceus* (*see* following pages for text).

Chapter 22

EASTERN WHIPBIRD

(Orthonychidae; Passeriformes)

Eastern Whipbirds are both common and sedentary, pairs remaining together to defend a territory throughout the year. Birds from the Atherton Tableland and adjacent highland areas generally occur above 300 metres except in extremely wet districts where they may descend to lower elevations. The breeding season is extended — July to January — but the bulk of activity takes place between October and December. A shallow cup of interwoven twigs is constructed close to the ground, usually in dense undergrowth. The eggs — two to three per clutch — are a lovely blue or blue-green with black squiggly markings. The female incubates the eggs and remains with the chicks until they leave the nest. During this period her mate brings food, feeding her as she remains on the nest. Both adults feed the young after they fledge. Juveniles and immatures are olive-brown, the latter showing only the palest indications of the white facial markings. These, and other adult colours, are not gained until the bird is one year old.

The closest Australian species to the Eastern Whipbird are the smaller Western Whipbird *P. nigrogularis* and the two species of wedgebills, all birds of the semi-arid and arid zones. The nearest relative of the whipbirds is unknown, but it may be a small, little-known species from highland New Guinea, the Green-backed Babbler *Androphobus viridis,* which looks like a miniature edition of the eastern bird. Some authorities include them with the logrunners and quail-thrushes while others place them nearest the whistlers or the babblers. Whatever the group is that they should most properly be allied to, the whipbirds are uniquely Australasian. The extraordinary song of the Eastern Whipbird *Psophodes olivaceus* is one of the best known sounds of the eastern forests — a long, clear, rising whistle, followed by two sharp notes. It sounds like a whip

being cracked, which has of course given the bird its name as well as several nicknames including Coachwhip Bird and Stockwhip Bird. This song is generally performed as a duet between the sexes. The male starts with the loud whistle, the concluding cracks being added by one female, sometimes more. Either sex may give its section of the song without a response from the other or produce the entire sequence itself.

Most people know the song far better than they know the bird because the Eastern Whipbirds are shy. But they are also curious and will eventually

Distribution
Eastern Whipbird.

emerge from the vegetation for a look, although brief, at a silent, stationary observer. These birds give the impression of flying with difficulty and are reluctant to do so, going only a short distance before diving back into the scrub. Unless startled or when crossing a clearing, whipbirds prefer to run. They move quickly and easily through the thick undergrowth of rainforest, wet eucalypt forest, dense gullies and other habitats. When disturbed, whipbirds react with a scolding call. While feeding, they noisily search through the leaf litter for insects, the sound of which is a clue to their presence. Using their bills, they flick and toss fallen debris aside.

Chapter 23

SCRUBWRENS

(Acanthizidae; Passeriformes)

Scrubwrens belong to a group of small birds which at a cursory glance appear drab and sombre but on closer inspection reveal a pleasant range of brown, fawn, buff and rufous coloured plumage and boldly-marked faces. Most are dwellers of the undergrowth in forested country along the coast and nearby ranges. Scrubwrens are widespread in New Guinea. Closely related forms live in low scrub throughout open country in other parts of Australia.

Yellow-throated Scrubwren

With its yellow throat and eyebrow and black facial mask, the Yellow-throated Scrubwren *Sericornis citreogularis* is the most colourful of this group of birds. It is also the largest. Females lack the pronounced black mask of the male. The long legs reflect the highly terrestrial nature of this bird; rarely does it move higher than a metre from the forest floor when feeding. Insects and small arthropods, supplemented with seeds, are picked up as the scrubwren forages through the leaf litter. The darkest parts of the forest are the favourite haunts of the Yellow-throated Scrubwren and its fondness for these gloomy places has earned it the name 'Devil Bird'. Of the scrubwrens, this species is the best songster. It calls with clear whistling notes interspersed with excellently imitated calls of other birds. Over 30 species have been recorded being mimicked.

The Yellow-throated Scrubwren is a common resident within its discontinuous range. In the north it is very much a bird of the highlands above 600 metres.

The pendulous nest of this species is a common sight in the rainforest, usually near streams. It is suspended from a branch, or placed in a tangle of vines and branches, within ten metres of the ground. Untidily constructed of leaves, twigs, roots and other coarse plant material, these nests may measure a metre from top to bottom. Feathers are used to line the inside, and mosses and lichens, and even small growing plants, are decoratively placed on the outside. This structure takes a pair of scrubwrens working together about two weeks to build. The eggs, usually three to a clutch, are chocolate brown encircled with darker bands. These take three weeks to hatch, incubated by both parents. In an additional three weeks the chicks are fledged and ready to leave the nest. More than one brood may be raised in a season, lasting from October to March. Many nests are built with up to three egg chambers; with each renesting another is lined and put into use.

Distribution Yellow-throated Scrubwren.

Large-billed Scrubwren

Often, however, the Yellow-throated Scrubwrens are ejected from the hard-built nests by their smaller relative, the Large-billed Scrubwren *S. magnirostris*. The intruding birds reline the nest chamber, covering up the eggs of the original builder. Large-billed Scrubwrens also employ the nests of other small birds, but often it is an instance of occupying a disused nest from a previous season rather than an active one usurped from its rightful owner. It is capable of building its own nest, a small round structure with a side entrance, but where the two scrubwrens co-exist, it shows a preference for the easy way out. It breeds from July to January, laying three to four speckled eggs.

The Large-billed Scrubwren differs from its Australian relatives by feeding high in the layers of

the forest, often in the company of a number of other species. It flits through the leaves and along the bark in search of small invertebrates. As it forages, it may briefly hover or hang upside down. This species is also the dullest of the Australian scrubwrens. There are no distinguishing marks, except for the long black bill. Indeed, the complete lack of any marking is the most identifiable feature of the Large-billed Scrubwren: it is probably the plainest of all Australian songbirds. Its voice is an undistiguished twittering and harsh chattering.

Atherton Scrubwren

The Large-billed Scrubwren is common throughout its range, though in the north it is more restricted to the highlands. It is there that the only problems of identification occur. Above 600 metres, in the Atherton Tableland, a second unmarked brown scrubwren is found. These birds are so similar that for many years no one realised the existence of two species. Slight differences had been noticed in the 1920s but were generally ignored; it was not until 1969 that the Atherton Scrubwren *S. keri* was 'rediscovered'. Further studies have revealed small but consistent differences. The Atherton Scrubwren is slightly larger with a darker face. These characteristics are not much help in the shadows, nor are the longer legs, but the latter point to more helpful behavioural distinctions. Birds that spend most of their time on the ground generally have proportionally longer legs than those that do not. Unlike its arboreal look-alike, the Atherton Scrubwren is a bird of the forest floor and undergrowth within two metres of the ground. It feeds more sedately, without the active flitting of the Large-billed Scrubwren.

Despite its recognition as a distinct species, the Atherton Scrubwren has not received much further

Distribution Fernwren and Atherton Scrubwren.

Opposite: Yellow-throated Scrubwrens *Sericornis citreogularis* — female (above); male (below).

A male White-browed Scrubwren *Sericornis frontalis*.

attention. Its voice is not fully-known and only a few nests have been found; these, and the eggs, are like those of the Large-billed Scrubwren.

White-browed Scrubwren

The best known of the scrubwrens, and probably one of the most familiar bush birds of eastern Australia, is the White-browed Scrubwren *S. frontalis*. Its range follows the coast from the Atherton Tableland district south to around Shark Bay in Western Australia and includes Tasmania. Throughout its distribution, this species shows marked geographic variation in its plumage and this has led to the establishment of separate names for these regional forms. The White-browed Scrubwren is at home in a variety of habitats as long as they provide thick undergrowth. In the east and in Tasmania this includes rainforest.

Females lack the well-defined facial pattern of males, the brown rather than black mask contrasting less with the pale throat and eyebrows. The birds illustrated are the typical form of the southeast. As one travels northwards, several of the characteristics change, though gradually: the underparts and parts of the facial pattern become increasingly a buffy-yellow and the mask becomes black, appearing much more pronounced. This form was long-known as the Buff-breasted Scrubwren and maintained as a separate species. In general appearance it resembles the Yellow-throated Scrubwren and distinguishing between them in the field can be a problem. One prominent 'give away' is the noticeably shorter legs of the buff-fronted White-browed Scrubwren. The Tasmanian subspecies is larger than the mainland birds and more uniformly brown with a less well

Distribution
White-browed
Scrubwren.

developed facial mask. It is sometimes called the Brown Scrubwren.

In habits these various forms of the White-browed Scrubwren are alike. The species is usually encountered in small noisy bands in the lower layers of the vegetation hunting for insects and seeds. They spend some time on the ground but less than the Yellow-throated Scrubwren. The first indication of this bird is often its harsh scolding chatter given from nearby cover when approached too closely for its liking. At first it gives the impression of shyness, but it soon becomes bolder and cannot resist investigating squeaking noises. It hops into view, scolding loudly. The pale eyes give this species a particularly glaring expression.

Breeding starts early in July, when the parents construct a bulky round nest, hidden in the undergrowth, under bark or in thick cover on the ground, and continues through January. The female normally lays two to three eggs per clutch. The White-browed Scrubwren shows little movement from the territory during the year other than a post-breeding dispersion of young birds.

The Brown Scrubwren shares the denser parts of the Tasmanian temperate rainforest with a closely related and smaller species, the Scrubtit *S. magnus.* These two birds have apparently evolved on separate occasions from the same mainland ancestor.

Tropical Scrubwren
The Australian range of every scrubwren extensively overlaps that of at least one other scrubwren species, with one exception — the Tropical Scrubwren *S. beccarii.* Though found throughout much of New Guinea, in Australia this species is limited to an area of the Cape York Peninsula and further south to just past Cooktown. It may marginally be in contact with the Large-billed Scrubwren in the southern limits but this is unconfirmed.

Its relationship to other scrubwrens is perplexing; in the north of Cape York Peninsula and in New Guinea the Tropical Scrubwren has white eyebrows and it is often considered the northern replacement for the White-browed Scrubwren. (The reddish iris prevents any misidentification). Yet it is a bird with wide plumage variation, even in one locality. In other parts of the Peninsula, most of the markings are reduced and birds have a cinnamon wash through the plumage, making them resemble the Large-billed Scrubwren. The Tropical Scrubwren, however, behaves differently in some ways to both these species. Its foraging zone, for instance, is wide, ranging from low undergrowth to the foliage of the lower canopy. Thus, it fills an intermediate position between the other two scrubwrens. Other features of its natural history, such as breeding behaviour, have not been investigated.

A name long in usage for this species was Little Scrubwren, which is now deemed inappropriate

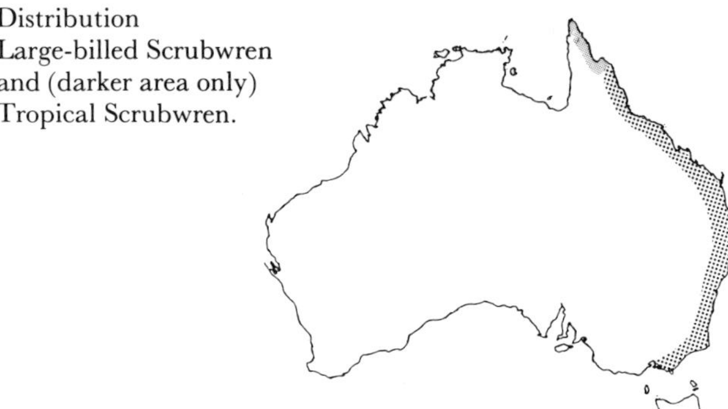

Distribution
Large-billed Scrubwren
and (darker area only)
Tropical Scrubwren.

because this bird is not significantly smaller than most other species. The name Tropical Scrubwren draws attention to its more northerly distribution (at least in Australia).

Fernwren
A near relative of the scrubwrens, which it resembles in appearance and much of its behaviour, is the

Large-billed Scrubwrens *Sericornis magnirostris.*

Robert Edden 1983.

Fernwren *Oreoscopus gutturalis*

Fernwren *Oreoscopus gutturalis*. Thought to belong to a small group of birds otherwise found in New Guinea, it is restricted to the highlands of the Atherton Tableland above 600 metres. Not an uncommon species, the Fernwren stays in the darker parts of the rainforest where it is not easily seen. It is not a shy bird, however, and will allow a quiet observer to watch it feed. It uses its feet to kick aside fallen leaves and twigs, exposing the small animals underneath. Now and then one has been seen following larger ground-feeding birds, like logrunners or scrubfowl, picking up items from the ground they disturb. Its range of calls include a high-pitched whistle and sharp chattering notes.

A major difference between the Fernwren and the scrubwrens is in their nesting habits. The general shape of the nest — a round dome with the entrance on the side — is similar. But rather than placing it in a thick tangle of vines or suspending it from a branch, the Fernwren secures its nest under a protruding bank or in a small cave. It lays two eggs; sometimes these are glossy white without markings, at other times they have a covering of small reddish specklings.

Chapter 24

THORNBILLS

(Acanthizidae; Passeriformes)

Brown Thornbill

The Brown Thornbill *Acanthiza pusilla* is one of a number of active little brown birds than can pose identification problems to the unwary observer. Often several species of thornbill occupy the same area, as well as a variety of warblers and wrens. They all travel together through the forest, actively foraging in the foliage. A patient observer can separate the different species by noting several plumage features which, in combination with observations of habitat, mode of feeding and calls, should lead to a confident identification. Inconspicuous but not shy, the Brown Thornbill pays little attention to humans, going about its task of finding food. A few squeaks, like a noisy kiss, will usually catch its attention and draw it nearer for a closer look. Identification of Brown Thornbills rests on the dark eye, streaked breast, scalloped forehead, reddish rump and unmarked sides of the face. These collective characters distinguish this species from all others.

This species is in fact one of the most common small birds of eastern Australia. It prefers moist

Distribution
Brown Thornbill
and (darker area only)
Mountain Thornbill.

vegetated areas with good ground cover. In addition to rainforest, the Brown Thornbill is found in eucalypt forest with thick undergrowth, mangroves, vegetation along watercourses, and freely enters suburban parks and gardens. Here single birds, pairs or family parties search restlessly for small insects, most gleaned from leaf surfaces; occasionally they hover to snatch an item off a more difficult site. Often they form mixed species feeding flocks. Where several species of thornbills are found together, differences in feeding zones reduces the competition for food. Brown Thornbills forage in the lower branches and in the understorey, whereas Striated Thornbills *A. lineata*, which overlap their distribution, and frequently enter temperate rainforest, tend to feed higher in the vegetation and more towards the outside of the tree.

Brown Thornbills are remarkably sedentary, moving only short distances from where they were hatched. One banded bird was recovered numerous times at the same location over a period of 13 years.

Brown Thornbill at nest.

Individuals this old are exceptional, most living only a few years.

Starting in June — later in colder regions — the thornbills build a nest of bark strips and grass in the shape of a dome. It has a side entrance, the top of which is usually shielded by a small hood. Egg sacs of spiders attached to the nest seems to be a regular feature. The nest is situated close to the ground and is not difficult to find with a little effort. Brown Thornbills lay three spotted eggs, though they are often parasitised by cuckoos.

An interesting characteristic of Brown Thornbills is their tendency to produce a soft, pleasant warble or to mimic the songs of other birds when under stress (such as being handled for banding). Their normal song is a 'pee-or' mixed with other notes; their alarm call a scolding buzz.

Mountain Thornbill

The Brown Thornbill goes as far north as the rainforests of mid-eastern Queensland, near Proserpine and Mackay. In the Atherton Tableland rainforest tract, it is replaced by the allied Mountain Thornbill *A. katherina*. For many years, this was thought to be a pale, poorly marked subspecies of the Brown Thornbill. But it was eventually recognised that it differed in several ways other than a more faintly patterned plumage. In the living bird, the most obvious distinction is the cream-white iris. In plumage and eye colour, the Mountain Thornbill resembles very young Brown Thornbills.

The Mountain Thornbill is common in rainforest in the highlands above 400 metres. Of the few nests that have been reported, they resembled those of the Brown Thornbills but were consistently larger and bulkier. The eggs, which have only recently been described, are like those of the Brown Thornbill. In the highlands of New Guinea is another related, white-eyed species, the Papuan Thornbill *A. murina*, the only thornbill found outside Australia.

Opposite: Brown Thornbill *Acanthiza pusilla* (above) and Brown Warbler *Gerygone mouki* (below).

Chapter 25

WARBLERS
(Acanthizidae; Passeriformes)

Gerygone warblers or fairy warblers are another group of confusing, yet ultimately recognisable small bush birds. They are often associated with thornbills and sometimes mistaken for them. But they lack the streaked breasts, patterned foreheads and coloured rumps of thornbills. Their bodies too are of a slighter build. Their voices are clearly more musical, as the name 'warbler' implies. A careful look is sufficient for most observers to differentiate between the two groups. Separation of the assorted species of warblers is based on patterns of the face and tail, in conjunction with habitat and calls.

Brown Warbler

Of the eight *Gerygone* warblers in Australia, the most widespread of the predominately rainforest species is the Brown Warbler *Gerygone mouki*. Three discrete populations occur along the east coast: in the Atherton Tableland district; the Clarke Range district; and from southern Queensland to northeastern Victoria. The northern subspecies lives at higher altitudes than those from elsewhere. Birds from each area are distinctive and at times the Brown Warbler has been divided into two species.

The Brown Warbler is a common resident of rainforest and humid sclerophyll forest with thick undergrowth. Its clear 'what-is-it', less elaborate than the voices of other warblers, is repeated throughout the day. A common component of the sound of the forest, it can be heard at all times of the year as ones, twos and small parties of these birds move actively among the leaves, feeding and calling. Most of their foraging takes place in the outer layers of the foliage, picking insects from the leaves or snapping them up in the air. Brown Warblers hover far more than thornbills. As they flutter in search of

Adult male Fairy (Black-throated) Warbler *Gerygone palpebrosa*.

prey, their spread tails reveal prominent white spots, another field mark of these birds.

Courtship begins in September when a pair starts to perform its display, fanning their tails while bowing to each other. Grass, moss and bark fibre, bound with cobweb and hung from a vine or thin branch, are woven together by the female. She bur-

Distribution
Brown Warbler.

Distribution
Fairy Warbler
(darker area only)
and Green-backed
Warbler.

rows into the side of this mass of plant matter and, by turning and pushing, hollows out the inside of the ball, creating the chamber in which she will deposit the eggs. She finishes by building up the sides of the entrance, extending it into a spout. A tail, some 70mm to 120mm long, hangs from the bottom. She lays and incubates three eggs which hatch in 12 to 14 days. In even less time — 10 to 12 days — the young have fledged and are ready to leave the nest. The short duration required to raise a brood to independence allows a pair of breeding birds to nest twice in a season. Nests of Brown Warblers are frequently parasitised by cuckoos or utilised by other species when abandoned.

Fairy Warbler

Northeastern rainforests have another common species of warbler, the Fairy Warbler *G. palpebrosa*. For years it was believed that there were two species of brown-backed, yellow-bellied warblers in this part of Queensland. The more southerly form has males which have the same white throats as females. Northern birds, sometimes called the Black-throated Warbler, have similar females but the males have dusky grey throats and a prominent white moustache stripe (this is the bird illustrated). It was then discovered that between Cooktown and the Atherton Tableland — the zone separating the two supposed species — males had brownish throats. This was interpreted as a transition between white-throated and black-throated forms. Because of this, all are now considered geographic variations of the one species. Other populations occur in New Guinea. The song of the Fairy Warbler is more melodious than that of the Brown Warbler but in breeding habits and foraging behaviour the species are similar.

In the northwest, the little known Green-backed Warbler *G. chloronota* inhabits monsoon forest, mangroves and rainforest. It is widely distributed in lowland New Guinea. What is known of its behaviour seems like that of related species. Most commonly found in mangroves, the Large-billed Warbler *G. magnirostris* also may be encountered in the thick scrub and gallery vegetation along waterways. It is perhaps the best songster of a group noted for their voices.

These birds have been known collectively by a variety of names: fairy warblers, fly-eaters, and most recently, gerygones. The last has been proposed as a replacement for the name 'warbler' on the grounds that the birds of the Australasian region are not related to the warblers of the Old World or to the wood warblers of the New World. However, the term 'warbler', as it applies to all these groups, has taken on a general meaning of small birds that feed primarily on insects taken from the surface of leaves and which have warbling calls. It does not seem worth the trouble to introduce a cumbersome name for a more descriptive one in common usage.

Chapter 26

LOVELY WREN

(Maluridae; Passeriformes)

Members of the chestnut-shouldered wrens, of which the variegated Wren *Malurus lamberti* is best known, replace each other across Australia. At least one form occupies most localities. Some of these are considered by ornithologists to be distinct species while others are judged to be merely subspecifically different. The status of the Cape York Peninsula representative, the Lovely Wren *M. amabilis,* remains to be settled conclusively but it is certainly one of the most distinctive members of this group.

The brilliantly coloured males differ only slightly in plumage characters from other populations. There is no problem in identification, however, for no other form is known to overlap it geographically. One of the most noticeable distinctions of the Lovely Wren is the plumage of the female. Except for the Lovely Wren and the birds restricted to Arnhem Land and the Kimberley Ranges, females of all other chestnut-shouldered wrens are a nondescript light brown. Those across northern Australia have grey-blue crowns and backs which contrast attractively with the rest of their plumage.

Lovely Wrens are also set apart from the others by their behaviour. Chestnut-shouldered wrens generally feed in low vegation, rarely moving more than a few metres from the ground. By comparison, Lovely Wrens spend much of their time high in trees, often up to 15 metres. Here they search for insects, moving through the foliage in family parties which sometimes comprise ten or more birds.

The principal habitats of the Lovely Wren are the edges of rainforest and along watercourses. They will also, at times, travel short distances into nearby open woodland. This wren is sedentary but each family group occupies a relatively large area throughout the year. The choice of section in this area varies from season to season.

Lovely Wrens may breed at any time conditions are suitable and up to two or three times in a year. The nest is a loosely woven dome with a side entrance, situated near the ground in a low shrub or clump of grass. With only a little help from her mate, the female constructs this in eight to ten days. She lays two to three finely speckled eggs which she incubates for 12 to 14 days. Once hatched, the chicks require another fortnight to gain their feathers. Like other species of these wrens, the Lovely Wren has 'helpers at the nest', non-breeding birds from earlier broods. Such males, which assist with the rearing of the young but do not breed themselves, retain a female-like plumage. Even the dominant male of the group may on occasion acquire a female-like plumage when not breeding.

The Lovely Wren and the other species of these Australasian birds are not related to the wrens of Europe and the New World but like them, they cock their tails and move actively through low vegetation; these similarities resulted in the same British name being applied. 'Fairy-wren' has been suggested as a substitute.

Distribution Lovely Wren.

Opposite: Lovely Wren *Malurus amabilis* — adult breeding male (above); non-breeding eclipse male (centre); adult female (below).

Chapter 27

WHITE-THROATED TREECREEPER

(Climacteridae; Passeriformes)

Virtually every part of the world has a bird which climbs up the sides of trees in search of insects. These are usually collectively known as creepers or treecreepers but are actually composed of several unrelated groups. In Australia, there are six species of treecreepers. Their relationship to other birds is obscure though it presumably lies closest to some group in this part of the world.

The common species of the rainforest is the White-throated Treecreeper *Climacteris leucophaea.* It has a disjunct distribution along the east coast while a very closely allied form (the Papuan Treecreeper *C. placens*) lives in the highlands of New Guinea. Those birds of southeastern Australia range through rainforest and wet eucalypt forest, to drier woodland and some inland scrubs. Northwards they become increasingly confined to rainforest and fringing forest. Additionally, treecreepers from the northern part of the range are more restricted to higher elevations: in the Atherton Tableland district, most are found above 300 metres. This northern form, once considered a separate species, is the Little Treecreeper. It has the grey throat and breast of the Treecreeper that dwells in the ranges of central-eastern Queensland. The birds illustrated are typical of the southern part of Australia.

All these forms of the White-throated Treecreeper are similar in voice and habits. The most frequently heard call is a series of sharp, piping notes which may descend or remain at the same pitch. These birds, as their name suggests, spend much of their day searching for insects in the bark of trees. Strong feet and claws help them climb the sides of trees as they probe for food in crevices, cracks and under bark with their long, curved bills. A variety of spiders and insects are eaten but ants are a particular delicacy. The White-throated Treecreeper flies to the base of a tree and begins to work up the side of the trunk, spiralling as it ascends. It moves upwards with a peculiar shuffling motion, always leading with the same foot. Unlike some other bark-feeding birds, such as sittellas, treecreepers cannot climb down head first. All their vertical movements are directed upwards, although they do have the ability to walk along the undersides of horizontal branches, hanging upside down. This species may opportunistically take nectar from flowers but unlike other Australian treecreepers which regularly forage on the ground, it rarely descends from the trees.

Sexes of the White-throated Treecreeper are alike except for the small chestnut mark near the ear of the female. This species remains in the same territory throughout the year. For most of this time, the male and female live and feed apart, reacting aggressively towards each other when they meet. They forge a temporary peace only when the breeding season starts. Their nest, which is a collection of debris, bark, dung and other scraps, is usually placed in a natural hollow in a trunk or occasionally at the end of a branch. The interior is lined with fur and feathers. A normal clutch is of two to three eggs, which are covered with reddish speckles. They are incubated by both parents, who also share the duties of rearing the young. Soon after the offspring have become reasonably independent, their parents' aggressive tendencies reassert themselves and the young birds are thrown out of the territory to fend for themselves.

Distribution White-throated Treecreeper.

Opposite: White-throated Treecreeper *Climacteris leucophaea* — male (left); female (right).

Robert Edden 1983

Chapter 28

OLIVE-BACKED SUNBIRD

(Nectariniidae; Passeriformes)

The brightly coloured, jewel-like sunbirds of Asia and Africa fill much of the same ecological role as do the hummingbirds of the New World tropics. Their flight is rapid and darting, frequently pausing to hover as they sip nectar from a flower. Of the 118 species, only one reaches Australia — the Olive-backed or Yellow-breasted Sunbird *Nectarinia jugularis*. And even this bird occurs widely outside the country, extending as far as southeast Asia and southern China as well as in New Guinea and the Solomon Islands.

Though the name Yellow-breasted Sunbird has long been in use it does not accurately describe the male: his iridescent black breast glimmers with flashes of blue and purple. It is only the females and immatures that have a yellow breast. These dainty acrobatic birds are a common sight in mid-eastern to northern Queensland, chasing each other for long periods or calling with their lively, high-pitched 'tzit-tzit-tzit'.

Associated with the edges rather than the depths of the rainforest, Olive-backed Sunbirds also inhabit vegetation along watercourses, mangroves, urban parks and gardens. Because they are tame and nest freely around settlements, these birds are familiar to most residents of these areas. The long, spindle-like nest is suspended from a vine or branch, or from a man-made object, such as a powerline or under the eaves of a house. The birds enter and leave through an opening on the side. Cocoons, bark, dead leaves, grass and other plant material are bound with spiderweb, and strings of caterpillar droppings are incorporated and often hang from the bottom of the nest, giving it a 'tail'. The inside of the nest chamber is lined with feathers and plant down. A favourite nest may be used for several seasons too — one recorded as having been re-employed for 15 years. Each season the owners make the necessary repairs until the accumulated wear and tear finally becomes too severe and the nest drops off its support.

Prior to the peak of breeding activity, males enact an unusual and enigmatic ritual. A band of individuals chase each other noisily for a period then suddenly halt; then sitting closely together, they pull their heads close to the body and point their bills upwards. This practice can last from several minutes to as long as an hour before it ends with the birds dispersing. What purpose this strange activity serves is not understood.

After the onset of breeding, a pair of sunbirds maintain a territory, guarding it vigorously, even against much larger birds. The male spends much of his time watching the female build the nest and incubate the eggs. It is not until a week after the young leave the nest that he starts to take an active part in their upbringing, assisting with the feeding. Sunbirds may raise three broods in a season, the first in August, the last finishing in January.

Distribution Olive-backed Sunbird.

Their long, slender, decurved bills helps sunbirds to probe for nectar, often while hovering in front of a flower. Larger or more difficult flowers are torn apart or pierced at the base. Nectar is by no means their only food. Insects are picked off leaves or captured in the air while spiders, a favourite food item, may be pulled from the webs by hovering birds.

Opposite: Olive-backed Sunbird *Nectarinia jugularis* — female (above); male (below).

Chapter 29

SILVEREYE AND MISTLETOEBIRD

(Zosteropidae; Dicaeidae; Passeriformes)

The migration patterns of Australian birds have yet to be thoroughly examined. A number of species are known to undertake regular annual movements but few have been the subject of any intensive investigation. Such studies involve trapping birds and marking them in some individual manner before freeing them to continue their flight, so that particular birds can be traced when recaptured. Accumulation of capture-recapture data of many individuals should build a picture of the seasonal movements of a species.

One of the species which has been given a sufficient degree of attention in this regard is the Grey-backed or Eastern Silvereye *Zosterops lateralis*. Each autumn thousands of these little birds depart on the northward leg of their periodic round trip. Much of the travelling is done at night and the following day a patch of trees or shrubs may be filled with Silvereyes, giving their pleasant warbles while they search for food to replenish the energy they have used before the next evening's journey. It is not known exactly how far they go before settling down for the winter. Clues to their place of origin are given by the intensity of brown on the flanks. Tasmanian breeding birds have a much richer and darker colour than those on the mainland and these make it at least as far as northern New South Wales. The further northward a bird originates the paler is its colour, but to confuse matters more, some individuals remain in their summer homes throughout winter. During the year, almost any locality in the south and central parts of eastern forests and woodlands, has a supply of Silvereyes. They are adaptable birds, which are equally at home in the rainforest or in a city garden.

Silvereyes are able to feed on a range of items. The short bill is well suited to gleaning insects from the foliage or picking berries from vines and shrubs. Rainforests offer good supplies of both. The tip of

Grey-backed Silvereye *Zosterops lateralis*.

the tongue, like that of honeyeaters, is fringed like a brush for collecting nectar from flowers. There are both good and bad points about the feeding habits of the Silvereye. This bird has developed a liking to fruits of both cultivated and unwanted exotic plants. It is known to poke a hole in the side of large fruits and hollow them out from within. Smaller fruits are punctured with a jab of the bill. It has in fact become such a pest to vineyards that some states will not protect this species under wildlife regulations. Silvereyes are also disseminators of the seeds of introduced weed species. On the other side of the ledger they eat large quantities of harmful insects.

This bird is a prolific breeder. Two or three clutches — each usually of three eggs — are raised in a year. The nest is a small cup with walls so thin that its sky-blue eggs can be seen within.

The Grey-backed Silvereye belongs to an exclusively Old World family which contains a number of very similar olive-green species. In countries outside Australia they are known collectively as 'white-eyes' or 'wax-eyes' for the small circle of white feathers around the eye. Our 'silvereye' seems too well established in Australia to change.

An able coloniser, the Grey-backed Silvereye breeds on Pacific islands as far east as Fiji. Australian birds have reached New Zealand, blown across the Tasman Sea, by prevailing winds which led, in the middle 1850s to a permanent colony being established. This is the last species to be given a Maori name, 'tauhou' meaning stranger, though the bird is now common in that country.

Another small bird which is confined to its food source rather than habitat is the Mistletoebird *Dicaeum hirundinaceum*. Home to this bird is anywhere that mistletoe berries are available, and they are a common feature of the rainforest flora. The Mistletoebird eats the berries, removing the fleshy outer coat, and deposits the seeds on a branch. Each seed has the potential to start another of these parasitic plants. These birds do not really have a stomach — an almost straight tube from mouth to vent is sufficient to remove the outer coating of the berries.

Distribution Grey-backed Silvereye.

Males are glossy blue-black above with a bright red throat and white underparts. Females are grey with a light red wash under the tail.

Chapter 30

HONEYEATERS

(Meliphagidae; Passeriformes)

Several families of birds are entirely restricted to the australasian region — or nearly so — but few are as characteristic of this part of the world or as widespread in it as the honeyeaters. Honeyeaters are distributed from eastern Indonesia and New Guinea through the South Pacific and New Zealand. They even reach Hawaii. In Australia, this family of birds is represented in virtually every part of the country, proving this group one of the most successful at exploiting almost every habitat on the continent.

This has been largely accomplished by the honeyeaters' eating habits. Australia has a great many species of flowering plants such as eucalypts, banksias, melaleucas, callistemons, etc., an array of nectar sources to which honeyeaters are well suited. A feature found in all species of these birds is a long, extendible tongue with a fringed brush-like tip, a design which allows honeyeaters to efficiently probe flowers for their nectar, collecting it in the feathery end. The edges of the tongue are folded over forming a tube, down which the nectar passes. At the same time, pollen collects on the forehead and face of the feeding honeyeaters which brushes off on other flowers as the birds make their rounds. This is why honeyeaters are among the most important pollinators of many native Australian plants — a mutually beneficial relationship. Indeed it is an essential relationship.

Not all species of honeyeater, however, exist solely on nectar. It ranges from almost 100 per cent for some species to only a small proportion for others. Alternative sources of sugary, nectar-like substances are the secretions of some insects and the juices exuded by plants at the sites of injury. Fruit is a common food item for some honeyeaters and the establishment of exotic fruit trees in gardens and orchards has been well received by these birds. An important part of the diet of some honeyeaters is insects, picked from flowers and leaves or actively pursued in flight.

Because flowering times vary among plant species, honeyeaters that depend upon nectar for a substantial portion of their diet must move to find continuing sources of food. This is the reason so many exhibit nomadic behaviour. Honeyeaters are normally noisy and aggressive but nowhere is this more obvious than at a favourite nectar-bearing tree when it is visited by a range of species. A considerable expenditure of energy goes towards chases and attempts at displacement of neighbours from choice flowers. There are several types of honeyeaters which feed along the edges of rainforest when trees are in blossom but that would not normally be considered rainforest species. The following text confines itself to those honeyeaters most characteristic of this habitat.

Members of this family are extremely diverse in size and form. The largest species is 450mm long, the smallest barely 100mm. Bills may be long, thin and decurved or relatively short and straight. Differences such as these help reduce competition between species because it permits them to partition food sources in various ways.

Lewin's, Lesser Lewin's and Graceful Honeyeaters

Despite the diversity shown by this family, some species stand out in many people's minds as 'typical' honeyeaters, none more so than the Lewin's Honeyeater *Meliphaga lewinii* (John Lewin, for whom it is named, was one of Australia's pioneering naturalists). This bird belongs to a compact group of about 13 very similar species with representatives in eastern Australia and New Guinea. Three of these occur in this country but two are restricted to northeastern Queensland. Lewin's Honeyeater is widespread and common in rainforest, wet eucalypt and adjacent open forest along coastal and highland

Opposite: Adult pair of Lewin's Honeyeaters *Meliphaga lewinii.*

regions. Never difficult to see, its boldness and curiosity allow observers to draw it close by making a loud kissing noise. A bird thus attracted will sit on a nearby branch peering and leaning forward to watch the observer. The distinctive staccato, machine gun-like chattering of the Lewin's Honeyeater is one of the most prominent calls of eastern wet forests.

A variety of foods are eaten by this honeyeater but it becomes more pugnacious than usual around a good source of nectar. Bouts of chasing take place as Lewin's Honeyeaters attempt to dislodge each other at flowers or to drive away smaller species of honeyeaters. Insects, secured from bark or leaves, also form an important part of the diet.

The Lewin's Honeyeater commences breeding somewhat earlier than most of the other honeyeaters (August) but otherwise its breeding biology is typical. Its nest is a cup-shaped construction of bark, moss, rootlets, twigs and plant fibre suspended by the rim from a horizontal fork. Like those of many other honeyeaters, the eggs are salmon-buff with scattered spots of darker red. Two to three eggs comprise the usual clutch. Incubation and fledging each require about two weeks.

In northeast Queensland, the Lewin's Honeyeater shares the rainforest with the other similar species, the Lesser Lewin's Honeyeater *M. notata* and the Graceful Honeyeater *M. gracilis*. The Lewin's Honeyeater occurs at all altitudes in the southern part of its range, though it rarely descends below 250 metres in the north. The other two species are generally birds of lowland rainforest but there is some overlap at intermediate elevations. Where all three co-exist they can present some problems of separation.

The Lewin's Honeyeater can be distinguished by its larger size and greyer face. If the birds are not clearly visible, calls provide good clues to field identification. The Lesser Lewin's Honeyeater has a harsh chattering — somewhat like that of a Lewin's

Distribution Lesser Lewin's Honeyeater.

Distribution Graceful Honeyeater.

Honeyeater but not as loud — plus other chirps and whistles. In contrast, the Graceful Honeyeater is a comparatively silent bird: its single clicking note is unlike the calls of the others.

The Lesser Lewin's Honeyeater is more closely allied to some New Guinea species than to the Lewin's Honeyeater and for this reason it has been suggested that its name is misleading. The usual alternative, Yellow-spotted Honeyeater, is no more appropriate, for it describes not only the other two Australian species but most of the New Guinea ones as well.

A noticeably smaller bird, the Graceful Honeyeater is well-named. Its slimmer body, thinner bill and more acrobatic foraging help to distinguish it from the Lesser Lewin's Honeyeater. Usually it hunts for food higher in the canopy but readily descends to feed among introduced Lantana bushes and in gardens. The Graceful Honeyeater is the only one of these species which also occurs in New Guinea.

Bridled and Eungella Honeyeaters

Another interesting group of honeyeaters consists of the Bridled Honeyeater *M. frenata* and its close relatives. The Bridled Honeyeater is primarily a bird of

Distribution Lewin's Honeyeater.

Opposite: Bridled Honeyeater *Meliphaga frenata* (above right); Lesser Lewin's Honeyeater *Meliphaga notata* (centre); Macleay's Honeyeater *Xanthotis macleayana* (below).

the highlands of the Atherton Tableland district (hence, its alternative name, Mountain Honeyeater) though there is evidence of a shift to the lowlands in winter.

It is the pattern of distribution of this group that is intriguing. In each of the major rainforest refuge areas within eastern Australia and New Guinea exists a closely related species, evidence that rainforest was once more widely spread than it is at present. The ancestor of this group presumably was spread through eastern Australia before being trapped in the various refuge areas. As well as having species in both highland and lowland New Guinea, this group is represented in Australia by: the Bridled Honeyeater; the Eungella Honeyeater *M. hindwoodi* of the Clarke Range district of central eastern Queensland; the Yellow-faced Honeyeater *M. chrysops*, a bird which has moved from rainforest into forests and woodlands of the east coast: and the White-lined Honeyeater *M. albilineata* of the Kimberleys of Western Australia and Arnhem Land in the Northern Territory where it spends much of its time in the rainforest gullies and pockets of the escarpments.

The comprehensive extent of this group's distribution was not realised until recently with the discovery of the Eungella Honeyeater. For many years it was thought that the Bridled Honeyeater extended from the Atherton Tableland to the Clarke Range. With its dark face, yellow and white bridle stripe and plumes, and bicoloured bill (the latter rare among honeyeaters) this bird is difficult to mistake for anything else. Yet, though an individual bird from the Clarke Range was photographed in 1959 and a specimen collected in 1975, it was not until 1976 that Wayne Longmore of the Australian Museum noticed differences in the Clarke Range birds that had been overlooked by everyone else — he realised that it was a distinct and unnamed species. Clarke Range birds are smaller and greyer than those from the Atherton Tableland, the

Distribution White-lined Honeyeater (in NW); Bridled Honeyeater (darker NE area); Eungella Honeyeater (black area to south).

plumage is streaked and the bill is black, not bicoloured. In 1983 this population was given a formal scientific name; it commemorates Keith Hindwood, one of Australia's most famous amateur ornithologists.

There is still much to learn about the Eungella Honeyeater. Sometimes its noisy, active behaviour as it congregates at a flowering tree makes it the most conspicuous bird of the local rainforest; at other times it seemingly is absent. When not feeding it has been observed sitting unobtrusively in a high tree. Whether the Eungella Honeyeater is overlooked at times or in fact undergoes movements to other areas is not yet known.

Macleay's and Tawny-breasted Honeyeaters

The bare yellow skin of the face, slender decurved bill and streaked plumage set the Macleay's Honeyeater *Xanthotis macleayana* apart from all other species. Named for the naturalist Sir William Macleay, this species differs from the previous ones

Distribution Tawny-breasted Honeyeater (darker, northern area only) and Macleay's Honeyeater.

by being a shyer, more silent bird as it moves through the vegetation in search of insects and spiders which make up the bulk of its diet. The hunting techniques used by Macleay's Honeyeater are sluggish in comparison to the active movements used by other species. Bark, clumps of vines and masses of dead leaves are searched by thrusting and probing into them with the bill. Nectar, including that from garden flowers, and small amounts of fruit are taken to supplement the diet.

Macleay's Honeyeater is not uncommon in its small range. It is a bird of the lowland rainforest, making odd forays into nearby mangroves. The breeding season is relatively short, October — December, and its nest and eggs resemble those of the Lewin's Honeyeater group. The call of the Macleay's Honeyeater is a five note 'too-wit too-wee-twit'.

Opposite: Eungella Honeyeater *Meliphaga hindwoodi*.

Robert Edden 1983.

Tawny-breasted Honeyeater *Xanthotis flaviventer.*

In the rainforests of northern Cape York Peninsula, the Macleay's Honeyeater is replaced by a large relative, the Tawny-breasted Honeyeater *X. flaviventer.* This bird lacks the heavily mottled plumage but shares the bare skin of the face and long bill. Widespread in New Guinea, the Tawny-breasted Honeyeater varies considerably in colour across its range. It is commonly found outside rainforest in mangroves and eucalypt forest where it opportunistically feeds on flowers, fruit or insects as they are encountered. It benefits from its large size which allows it to chase away the smaller honeyeaters which attempt to share a food source. Unlike its quiet relative, the Macleay's Honeyeater, the Tawny-breasted Honeyeater is vocal and aggressive and may be encountered in noisy feeding parties.

Scarlet and Dusky Honeyeaters

Quite different in appearance from those species introduced thus far, the dainty Scarlet Honeyeater *Myzomela sanguinolenta* is one of the smallest honeyeaters. The genus of birds to which it belongs are all similar in form and have slender, sickle-shaped bills.

Most of these species, the only group of honeyeaters with prominent red in the plumage, have pronounced sexual dimorphism, a feature rare elsewhere in the family. Males are the brighter of the sexes and may have bright patches of red, contrasting with the rest of their brown or black plumage. Such is the case with the Scarlet Honeyeater. The brilliant male is red over most of his body, a feature which has earned the nickname 'Blood Bird'. His mate is a dull brown, adorned with only faint traces of red.

Their coloration and more conspicuous behaviour makes the males far more obvious than the females. Highly dependent on nectar, Scarlet Honeyeaters are nomadic within their range, following the appearance of flowering trees. They are not particular about where these occur and may turn up in rainforest, eucalypt forest, swamps, heaths or cultivated trees in streets or gardens if suitable flowers are available. Eucalypts in forests or along rainforest edges are particularly favoured. At these times, the clear tinkling song is commonly heard, usually high in the tops of trees, for Scarlet Honeyeaters are birds of the upper canopy surface. The female is more retiring and feeds away from her mate, in a more restrained manner. The comparative brightness and extroverted behaviour of the male make him the far more obvious sex, giving the impression of a very unbalanced sex ratio.

The larger Dusky Honeyeater *M. obscura* has the same shape as the Scarlet Honeyeater but lacks the marked sexual differences in colour, the red in its plumage limited to a wash through otherwise uniform brown feathers. This is a more tropical species than its relative, ranging to New Guinea, the Moluccas and northern Australia. It has a noticeable liking for eucalypt woodlands but also frequents a number of other habitats, including

Distribution
Scarlet
Honeyeater.

Opposite: Scarlet Honeyeater *Myzomela sanguinolenta* adult male (above); adult female (below).

Robert Felden 1983.

rainforest. The Dusky Honeyeater feeds in the crowns of trees but is easily lured close to an observer with whistles and squeaks.

Eastern Spinebill

One of the best known birds of suburban parks and gardens is the Eastern Spinebill *Acanthorhynchus tenuirostris*. It is an active honeyeater that pursues flying insects energetically, hovers as it probes flowers for nectar, and engages in animated chases with individuals. Although the Eastern Spinebill has readily adapted to civilisation, it is by no means restricted to the vicinity of human settlement. This bird is at home in wet eucalypt forests and heaths and along rainforest edges, its presence largely dictated by the availability of blossom-bearing plants. The long bill allows the spinebill to extract nectar from tubular flowers that is beyond the reach of

A male Eastern Spinebill *Acanthorynchus tenuirostris.*

other species. Flight is swift but two features confirm this bird's identification as it darts quickly past: prominent flashes of white in the spread tail as it flies and an audible clapping made by the rapidly beating wings.

Green-backed Honeyeater

Like most birds whose Australian range is confined to the far north of Cape York Peninsula, the Green-backed Honeyeater *Glycichaera fallax* is not a well-known species. Also found in New Guinea, this species was not recorded in Australia until 1913. Neither the egg nor the nest have yet been described. The bird itself has few identifying features. Its rather nondescript olive-grey plumage gives it a resemblance to the Silvereye. Observations of the Green-backed Honeyeater when feeding indicate that it spends much of its time actively gathering insects which form the greatest part of its diet. The short bill is not well suited for feeding on nectar.

Helmeted Friarbird

The Helmeted Friarbird *Philemon buceroides* is one of Australia's largest honeyeaters, belonging to a distinctive genus of birds with partially or totally naked heads, and usually a large knob at the base of the upper bill. Geographically separated populations of this species occupy different habitats; Arnhem Land birds are found in the rainforest gullies of the escarpment, while in north Queensland this friarbird frequently, though not exclusively, feeds along the edge of rainforests. At a flowering tree, the aggressive Helmeted Friarbird may make use of its large size to dominate and drive off other birds. Its large, pale buff eggs with brown and lavender blotches are unlike those of the smaller honeyeaters.

Several other species of honeyeater could justifiably be considered rainforest birds. In Tasmania, the Crescent Honeyeater *Phylidonyris pyrrhoptera* and Strong-billed Honeyeater *Melithreptus*

Helmeted Friarbird *Philemon buceroides*.

Distribution
Helmeted
Friarbird.

validirostris are common in the temperate rainforest.
The latter fills the gap of the treecreepers, a group of
birds absent on the island, clambering up the trunks
of trees after food. The White-streaked Honeyeater
Trichodere cockerelli, interesting as the only bird endemic to far northern Cape York Peninsula, is
mainly an inhabitant of heathland and woodland,
but it can also be found at times along the edges of
rainforest.

Chapter 31

FINCHES

(Estrildidae; Passeriformes)

Finches conical bill are an adaptation for cracking and eating seeds, a development that apparently has evolved on several independent occasions. Of the different groups of birds sharing the general name 'finch', only one, the estrildid finches (grassfinches and mannikins) occurs naturally in Australia. The majority of the 18 attractively coloured and patterned native species are inhabitants of open country. The only two species associated with rainforests are birds of the edges and clearings, rather than of the interior. One is widespread in Australia, the other very restricted.

Blue-faced Finch

Though a common species in eastern Indonesia, through New Guinea and the Solomon Islands to Vanuatu and several other South Pacific islands, the Blue-faced Finch *Erythrura trichroa* is a bird of scattered and enigmatic occurrence in Australia. Until the last few years, there were no more than 12 records of this species in the country. It appeared irregularly without any discernible pattern, but recently there have been almost as many sightings again, though for no apparent reason. In its extralimital range, the Blue-faced Finch occupies a variety of habitats, many quite open. In Australia, most records come from the fringes of rainforests and mangroves from sea level to high altitude. It is generally seen feeding on the ground or in low vegetation.

This species is very social for much of the year, forming flocks of 20 to 30 birds, but it pairs up for breeding. Like other estrildid finches, the Blue-faced Finch has an interesting courtship ritual. A courting male approaches a prospective mate holding a piece of grass in his bill. He gives a trilling call while angling his tail towards the female. She adopts a similar posture and both bob up and down. She will then fly off, pursued by the male. They land and repeat the sequence again, sometimes continuing 20 times or more before she accepts him as her mate.

Together they construct a bulky nest of grass and stringy vegetation. The male brings material to the female who fits it into the nest structure. When finished, an entrance tunnel 30mm long protrudes at an angle from one side. The female incubates three to five white eggs during the day; at night both parents roost inside the nest. This arrangement continues for 12 to 14 days until the eggs are incubated. The young fledge 21 days after hatching, fed actively by both adults. Newly hatched estrildid finches have distinctive, species-specific patterns of black markings on the inside of their mouths. In addition, the Blue-faced Finch chicks have small, iridescent blue 'beads' at the corners of the gape which disappear as the birds mature. They may help the parents to feed the young in the darkness of the nest chamber. Immature Blue-faced Finches lack the blue coloration of the adults. Very few records of breeding exist in Australia though nests have been found in March and November.

This species belongs to a genus known collectively as 'parrot finches'. All have green bodies but vary in the colour of the face which may be red, blue or multicoloured. The Blue-faced Finch is a popular aviary bird in Australia.

Opposite: Red-browed Finch *Aegintha temporalis* (above); and Blue-faced Finch *Erythrura trichroa* (below).

Red-browed Finch

Far better known to most people is the Red-browed Finch *Aegintha temporalis*. This species occurs all down the east coast where it is common to abundant anywhere there is sufficient ground cover to provide shelter, be it in human settlement or the country. It has adapted well to partial clearing of rainforest, finding the thick regrowth of invading species (Lantana, Blackberry, etc) much to its liking. The Red-browed Finch is highly social and the most confiding of the estrildid finches in the presence of humans. Flocks of these finches feed on grassy areas on the ground, never venturing too far away from thick vegetation. If disturbed they make a hasty retreat to its cover.

Active nests have been found in the southeast in every month of the year except July and August. In the northern part of this species' range, breeding is confined to January-April, the last half of the wet season. The nest resembles that of the Blue-faced Finch but the entrance tube is only half the length. Parents alternate in incubation duties during the day and roost together at night. Juveniles, which join the adults in large flocks, lack the red eyebrows

Distribution Red-browed Finch.

and have black bills. Across the northern half of Cape York Peninsula, Red-browed Finches are smaller and brighter birds than those to the south, with considerably whiter underparts.

Chapter 32

ORIOLES AND FIGBIRDS

(Oriolidae; Passeriformes)

Orioles are a widespread Old World family of fruit-eating birds. Many are boldly marked with yellow and black but the two Australian species are more modestly patterned. Close relatives of the orioles are the figbirds, confined to Australia, New Guinea and some eastern Indonesian islands. All enter rainforest as well as other forest types, their distributions being largely regulated by the presence of suitable supplies of fruit. Wholly arboreal, orioles and figbirds may feed together in the canopy along with other frugivorous species. Being aggressive, they attempt to chase other birds from a fruit-bearing tree. The ranges of the three Australian species overlap in northern sections of Queensland, Western Australia and the Northern Territory.

Yellow and Olive-backed Orioles

The most restricted of these species is the Yellow Oriole *Oriolus flavotinctus.* Confined to the high tropics, it does not extend south past the Mackay region of Queensland. This oriole is common in riverside trees, never venturing far from the vicinity of water where it sits producing its loud bubbling song. A feature of the northern forests throughout the year, the rich 'yok yok yoddle' may be repeated incessantly for 30 minutes or more. The Yellow Oriole is also an excellent mimic. Like many fruit-eating birds, this species is nomadic, following the appearance of ripening fruit. Females and immatures are brighter yellow than the male (illustrated) and are more heavily streaked.

The more widespread, but less colourful Olive-backed Oriole *O. sagittatus* shows similar nomadic tendencies in the northern part of its Australian range. Southern birds are migratory, departing their breeding ground in May to join or 'leap frog' northern populations. Some cross Torres Strait to New Guinea. They return in August to initiate breeding.

Both orioles build deep cup-shaped nests which they hang from a fork in the outer branches of a tree.

Two eggs (sometimes up to four in the Olive-backed Oriole) are laid.

As well as having a wider distribution than its relative, the Olive-backed Oriole has a greater range of habitats. From the humid coastal districts, it extends over the Dividing Ranges into drier regions of the west, including mallee. Native fruit is not prolific in these areas, other than cultivated crops, so orioles may become agricultural pests, particularly of vineyards. Insects form a large part of the diet when fruit is unavailable. These may be gleaned from the foliage or captured in mid-air.

The song of the Olive-backed Oriole is a somewhat monotonous 'orry orry ooo', less resonant than that of the Yellow Oriole.

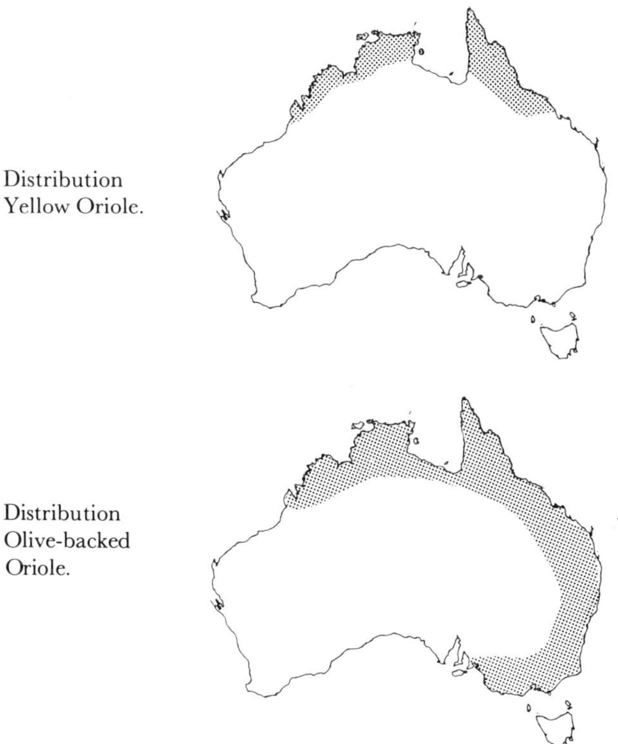

Distribution Yellow Oriole.

Distribution Olive-backed Oriole.

Figbird

Female Figbirds *Sphecotheres viridis,* and immatures of both sexes, may initially be mistaken for orioles

but they have black rather than pink bills and they are much browner and more heavily streaked. Adult males differ in colour (black top of head, olive-green back) and by having a bare patch of reddish skin around the eye. This patch becomes brighter when the bird is excited or annoyed. It remains coloured during the breeding season. There are two distinct forms of the Figbird in Australia, formerly considered separate species but which are now known to interbreed in mid-coastal Queensland. In the northern form, males are bright yellow underneath; this has been called the Yellow Figbird. The other, the Southern Figbird, has a grey neck and throat and light olive-green underparts.

Figbirds occur along the edges of rainforest, avoiding the denser interior, and in watercourse vegetation, orchards and gardens. Parks where Moreton Bay Figs have been planted may support colonies of Figbirds once the trees are mature enough to produce fruit. The arrival of Figbirds in Hyde Park, Sydney, was documented. They became resident there in 1946 and are now a common sight in the large fig trees. Active flocks of these birds are seen moving from tree to tree or heard giving their squeaky 'cheyr' or whistling 'tu-heer tu-heer'. As they feed, they may dangle head down to reach fruit at the ends of small branches. A similar position has been observed during rainstorms, when figbirds actually hang upside down below the branch, apparently bathing. This odd pose may be held for over a minute.

Flocks of immatures gather to wander after

Yellow Figbird *Sphecotheres viridis flaviventris* — female (left); male (right).

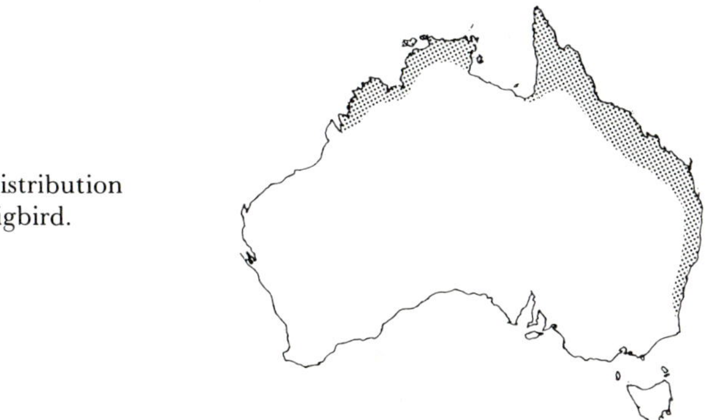

Distribution Figbird.

breeding. Adults may join these flocks or remain in the general vicinity of their nesting territories.

Opposite: Yellow Oriole *Oriolus flavotinctus*.

Chapter 33

METALLIC STARLING

(Sturnidae; Passeriformes)

A colony of Metallic or Shining Starlings *Aplonis metallica* is a spectacular sight as hundreds to thousands of birds suspend their globular hanging nests in a large tree, often joining more than one together. Such sites become well known for they are used year after year. Adults, joined later by growing young, visit and leave the colony throughout the day. During the nesting period, broken eggs and dead chicks fall to the ground, attracting a range of scavengers.

These starlings are abundant, gathering in huge flocks, which are a common sight through their range, particularly after breeding. Though centred on rainforest and fringing habitats, Metallic Starlings have now become so tolerant of man that large numbers roost in trees along city streets and on occasion, even form nesting colonies within city limits. Their entire mode of behaviour is much different than that exhibited by the Common Starling *Sturnus vulgaris* and Indian Myna *Acridotheres tristis,* two introduced starlings which have made themselves unwanted guests around human habitation.

The Metallic Starling has not become an urban pest like the other two species, nor is it a scavenger for man's scraps, fruit making up to 95 per cent of its diet. The only area in which it comes into conflict with man is when it attacks cultivated fruits, with entire flocks converging on a fruiting tree amid considerable calling and commotion. They clamber through the branches in their search, knocking as much fruit to the ground as they manage to eat.

A flock contains a mixture of adult and immature birds, the former easily distinguished by the glossy black plumage with green and purple iridescence, and long pointed nape feathers. Immatures are glossy black on the back with white underparts, streaked with black and are capable of breeding while in this plumage. The large bulging

Shining Starling *Aplonis metallica* — immature (left); adult (right).

Distribution Shining Starling.

orange-red eyes of these starlings contrast dramatically with their black feathers.

A pair may raise two to three broods in a season, laying two to four eggs each time. After raising their young, they migrate to New Guinea in February — April. They return to Queensland in the following August — September to repeat the cycle. Some individuals are resident throughout the year. This species is also found in the Solomon Islands and the Moluccas.

Chapter 34

SPANGLED DRONGO

(Dicruridae; Passeriformes)

'Drongo' in Australian slang means a fool. It was not, however, derived from the bird, but a particularly hapless racehorse early in this century. Whether the horse was named after the feathered drongo is another question. 'Drongo' in fact is the native Madagascar name for one of the African species of these birds.

The drongo family is comprised of 20 species and it extends from Africa through southern Asia to Australia. Nevertheless only one species is found here, the Spangled Drongo *Dicrurus hottentottus*. It is a species that is widespread elsewhere, occuring in India, China, the Philippines, south east Asia, Indonesia and New Guinea. Each population, particularly those on islands, shows a different arrangement of the hair-like feathers on the back of the head. This adornment is not as elaborate on the Australian birds as those of some of the other localities but they are nonetheless striking in appearance. Adults are glossy black with iridescent green spots (the spangles) on the head, neck and breast. The iris is dark red and prominent bristles surround the hooked beak. However, the most noticeable feature is its long, forked, fish-like tail. This diagnostic character makes it impossible to confuse the Spangled Drongo with any other bird in Australia.

Distribution
Spangled Drongo.

Spangled Drongos are difficult to overlook when they are around since they tend to sit on a conspicuous perch, often a dead limb in the top of a tree. From this vantage point the bird searches for flying insects, springing from its perch to pursue its intended victim with acrobatic but erratic flight. Swooping and diving, the drongo returns to a branch to eat its capture. Even if not seen first, the Spangled Drongo quickly draws attention to itself by its loud, strange calls. It utters harsh chatterings and a bizarre metallic note which has been likened to twanging a tautly drawn fence wire.

This species is bold and aggressive. Reports show that it has killed small birds on occasion. Nor is it shy around humans when residing from time to time in parks and gardens. Drongos occur in such non-rainforest habitats as mangroves and open woodland, but display a strong affinity for lowland rainforest. They avoid the thicker interior under the closed canopy, preferring instead the forest edge.

It is in this location that the Spangled Drongo chooses to nest. A delicate cup of twigs and vine tendrils, bound together with spiderwebs, is hung between a horizontal fork. The nest site is usually selected in the outer foliage of a tree, 10 metres to 20 metres above the ground. Three to five eggs are laid in September or October. These are pinkish or mauve, marked with darker streaks and spots of the same colours. The breeding season continues until March when many drongos migrate northwards. Those from the north of Western Australia and the Northern Territory cross the Timor Sea to Indonesia while eastern Australian birds move to New Guinea across the Torres Strait.

Some drongos exhibit an anomalous behaviour at this time. Instead of heading north, they remain through the winter in the same area — and some even move southwards. Spangled Drongos breed as far south as northeastern New South Wales but during winter they regularly appear around Sydney. They have even — on rare occasions — turned up in Tasmania. The significance or cause of these mixed patterns of annual movement is not known.

The relationships of the drongos are obscure. Recent studies suggest that they may be closest to the monarch flycatchers. Drongos' flycatching habits are suggestive of such connections but this may be due to convergence.

Adult pair of Spangled Drongos *Dicrurus hottentottus.*

Robert Seldon 1992

Robert Elden 1983

Chapter 35

BIRDS OF PARADISE

(Paradiseidae; Passeriformes)

Of all the families of birds in the world, few rival the birds of paradise for their dazzling colours and decorative feathers. Yet despite their international attraction they are found in only a small part of the world. Of the 43 species of birds of paradise, all but six are restricted to New Guinea and its satellite islands. The others occur in the Moluccas (two species) and eastern Australia (four species — two of which are shared with New Guinea). Surprisingly, these birds have not reached the Bismark Archipelago or the Solomon Islands, only relatively short distances from mainland New Guinea. The most spectacularly ornamental species found to the north have never found their way here.

Characteristically the males bear the elaborate plumes and glossy plumage; females are brown with streaks and mottlings which, though striking patterns in themselves, they are in no way comparable to those of the males. This disparity between the sexes is a reflection of their breeding system. Males are promiscuous, mating with as many females as possible. So their gaudy plumage has evolved to attract mates, a result of competition between males. Arboreal displays by males, often in a communal tree, involve an exaggerated exhibition of their most ornate feathers. The pair bond for only a matter of minutes, after which each female is left on her own. She must build the nest and raise the young without any assistance from the male. Only a few species, including one in Australia, are monogamous. These lack obvious sexual dimorphism and both sexes share the duties of rearing the family.

Riflebirds

Riflebirds are so-named because their call has been likened to the discharge of an old blunderbuss. The Paradise Riflebird *Ptiloris paradiseus* (shown here) of southern Queensland and northern New South Wales, and Victoria's Riflebird *P. victoriae* (named for Queen Victoria) of the Atherton Tableland rainforest tract, are both very similar in their habits. Some ornithologists consider them to be the same species. The northern bird is smaller and has a shorter bill, and has small plumage differences. Brown birds (females and immatures) are most often seen. The black plumage of the adult male does not start to appear until the third year and it is not complete for another year.

Both species occur in rainforest and adjacent wet eucalypt forest, particularly above 500 metres. They clamber around the trunks of trees like large treecreepers, probing in crevices and under bark for insects, though the bulk of their diet is fruit. Riflebirds are very acrobatic when feeding, hanging upside down to reach berries at the end of branches. The bill is used to probe out the soft flesh from larger fruit or to remove its peel. Small parties of riflebirds collect at fruit-bearing trees, chasing off frugivorous birds of other species if they can.

The breeding seasons of the riflebirds is much like that of other birds of paradise in that it involves calling by the male and his complex courtship ceremony. He uses his drawn out, explosive 'yaa-aa-as' every few minutes throughout the day to advertise his presence to females and announce his territory to other males. Unlike many birds, he does not defend a breeding territory around the nest, Instead he guards a display territory centred on a prominent perch — such as a tree stump or exposed horizontal branch — where he performs his display to females.

Spreading his broad, rounded wings to their full extent, he cocks his tail over his back and bows forward. Throwing his head back, he points his bill skywards and spreads the metallic blue-green bib and elongate flank feathers. This is accompanied by

Opposite: Paradise Riflebird *Ptiloris paradiseus* — female (above); male (below).

grating calls. The wings, still fanned, are thrown forward so that they touch in front of the body. The inside of the mouth, shown when the male opens his bill, is a brilliant lime green colour which contrasts strikingly with his black plumages as he sways his head from side to side. Each movement of the ritual is carried out with a jerky rhythm. As the male performs, brown birds gather to watch, behaving in an excited manner. When a female lands next to him, he envelopes her with his wings, clapping them up and down as he moves around her. Young males which have not yet acquired the complete adult plumage also display and are occasionally successful in attracting a mate.

The female builds a bulky cup of stems and broad leaves, bound with vine tendrils and lined with smaller leaves and twigs. She further decorates the rim of the nest with an old snake skin, or nowadays, a man-made equivalent, a piece of cellophane. Moss and orchids are used as further adornments. The nest is situated in a tangle of vines or branches five to 30 metres from the ground. The female deposits two eggs, distinctively marked with bold reddish and mauve strips overlaid on a pinkish buff background. Incubation requires 15 to 16 days, and fledging another four weeks. Breeding takes place from September to January. The female raises a single brood per season. For the rest of the year, riflebirds are sedentary and quiet; at this time, a distinctive noise made by the wings of the male in flight is heard more often than his call. This sound has been described as like rustling silk or tissue paper.

Victoria's Riflebird is more common than the Paradise Riflebird. It has not suffered the depletions associated with habitat loss that the southern species has.

Magnificent Riflebird

Through the lowlands of New Guinea and along the coastal areas of northern Cape York Peninsula lives the third and largest species of riflebird, the appropriately named Magnificent Riflebird *P. magnificus*. It is also called Prince Albert's Riflebird. The broad blue bib of the male is bordered below by a band of bronze green, the flank feathers are much longer and more ornate than those of the other species, and the feathers of the belly lack iridescent edging. The female Magnificent Riflebird has redder upperparts than the smaller species, and white underparts, thinly barred with black.

An adult Trumpet Manucode *Manucodia keraudrenii*.

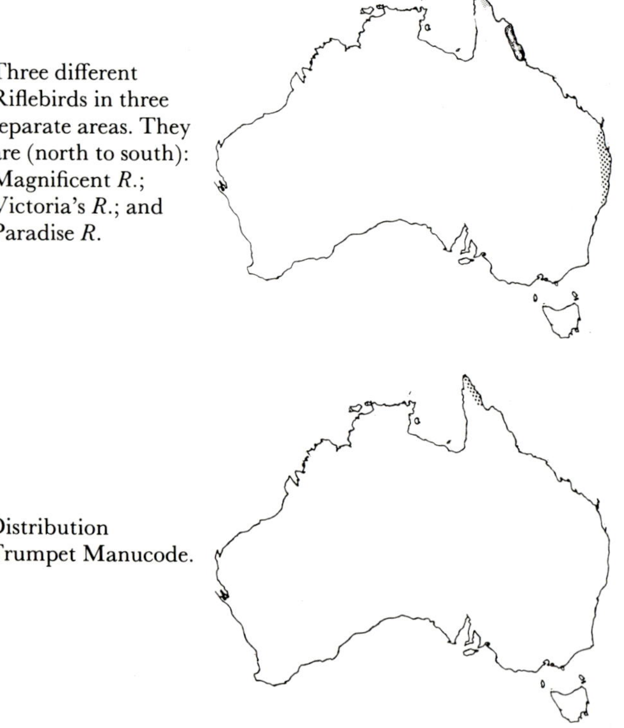

Three different Riflebirds in three separate areas. They are (north to south): Magnificent *R.*; Victoria's *R.*; and Paradise *R.*

Distribution Trumpet Manucode.

On Cape York Peninsula, this species is common to abundant in rainforest, monsoon forest and gallery vegetation, though as it feeds in the upper levels of the forest, it may not be noticed until it calls. In voice it differs from the other riflebirds by having three loud whistles followed by an extended one, 'whee-oo whee-oo whee-oo who-oo', given by both sexes. Males can be identified by their louder voices. The main courtship display of the Magnificent Riflebird is similar to that of its relatives, but apparently lacks the calling and open mouth elements. It also has other, less elaborate displays. The major aspects of the biology of the Magnificent Riflebird resemble those of the southern riflebirds.

Trumpet Manucode

The Trumpet Manucode *Manucodia keraundrenii* is unique among the Australian birds of paradise in having similar males and females which form monogamous pairs. The male still has an interesting ritual but this is performed to his mate within their breeding territory, not from an area specifically devoted to displays only, nor to any female which comes near. He pursues his potential mate through the forest, punctuated by period breaks when they perch and face each other. The male erects his feathers, spreads his wings, gives a single loud cry and then the chase starts again. The nest is a shallow saucer made of small vines and broad leaves, placed high in a tree. Its walls are thin and the two eggs can sometimes be seen within. Both sexes of Trumpet Manucode incubate the eggs and feed the young.

This species, also called the Trumpetbird, is named for its call, a single loud note, described variously as an organ note or a squeaking door. This is produced in the specialised windpipe or trachea. Additional loops of the windpipe are curled between the skin and muscles of the chest, this extra length acting as an organ pipe for the production of the manucode's call. This sound is apparently given only by the male; the female lacks the lengthened windpipe.

Four species of manucode, including the Trumpet Manucode, inhabit New Guinea. The other three are considerably larger and have more ornate, textured plumage; they share the modified windpipe. In Australia, the Trumpet Manucode occupies lowland rainforest and eucalypt forest. There is apparently some migration to New Guinea by at least part of the Australian population but this facet of the manucode's natural history is not well understood. It is difficult to assess the abundance of the Trumpet Manucode because it is shy and furtive. Most of its time is spent in the upper levels of the forest, seeking out the fruit which makes up the majority of its diet.

Chapter 36

BOWERBIRDS

(Ptilonorhynchidae; Passeriformes)

If it were not for their peculiar breeding biology, bowerbirds would be considered attractive, though not particularly noteworthy. In reality, of course, their courting behaviour has captured the fancy of many and intrigued scientists. Though presumably related to the birds of paradise, bowerbirds lack the elaborate and specialised plumes which have become so dominant in the courtship rituals of that family. Instead, the focal point of the male's courtship activities is a complex structure — the bower — which he builds himself. Its construction and defense occupies much of his time. It is here that he displays to potential mates, copulating with them when they enter the bower. Bowerbirds are often markedly sexually dimorphic, males being the more brightly coloured. As a generalisation, those species with the more colourful males have the less elaborate bowers — at least when considering the species in the drier parts of Australia and in the rainforests and savannahs of New Guinea. Another general rule is that when there is a great difference between the sexes that species is invariably polygamous (one male having many mates), in which case the pair bond is short and the roles of the sexes in breeding sharply divided.

Three major types of bowers are recognised. The simplest is the 'stage', a cleared area on the forest floor, usually decorated with leaves. An 'avenue' consists of two parallel walls of interlaced sticks attached to a broad flat platform, also woven of sticks. The largest bowers are the 'maypoles', these are structures of sticks built around the base of one or more saplings, often to a height of several metres. Representatives of all these types of bower builders can be found in the bowerbirds of Australian rainforests.

Golden Bowerbird

The only maypole builder in this country is the Golden Bowerbird *Prionodura newtoniana*, which dwells above 900 metres in the mountain rainforests of the Atherton Tableland district. Its maypole is a double structure extending up the sides of two small trees about a metre apart. A criterion for selecting a site for the bower is that these trees must have a branch or fallen limb between them, roughly parallel to the ground. This serves as a perch from which the male displays. Sticks are placed around each sapling, gradually forming a sheath surrounding the base. Bowers are reused each year with additional material continually being added. They grow in height, sometimes to three metres, and begin to spread out at the bottom. Eventually these two mounds meet, forming the characteristic double maypole of this species. White flowers are often placed at the ends of the display perch. Berries and lichen are also used for this purpose, though less frequently.

The male Golden Bowerbird performs on and around his bower, in the hope of attracting females. In a display not known in any other bowerbird, he hovers near the bower, fanning his tail. Another of his ceremonies is given from the display perch — he rocks from side to side, spreading the tail, fluttering the wings and erecting his short crest. Strange chatterings, which often sound mechanical, are uttered during breeding activities.

The male copulates with females attracted into the bower, then remains with the bower while they retire to build nests and raise young. Males mate with as many females as possible but they take no part in any subsequent nesting duties. A number of males form loose clans but each guards his individual territory against the intrusion of others. On the other hand, females and young males with immature plumage are allowed in the territory where

Opposite: Golden Bowerbirds, *Prionodura newtoniana* adult female (above); adult male (below).

they sometimes add to the bower or carry out mock displays themselves. Males require three years to obtain the bright yellow plumage of the adult.

Like most bowerbirds, the Golden Bowerbird largely eats fruit. But it has been observed hovering around leaves, possibly to pick insects from their surfaces.

Bower of the Golden Bowerbird.

Satin Bowerbird

In the Australian rainforests, the avenue builders are represented by the Satin Bowerbird *Ptilonorhynchus violaceus* and Regent Bowerbird *Sericulus chrysocephalus*. The former is by far the best-known of all the bowerbirds and is the subject of many studies. Because the Satin Bowerbird is common through most of its range, and its bowers are easy to locate, excellent photographic records have been made of its courtship behaviour.

The adult male is a glossy blue-black with a purple sheen, its light blue eyes striking in contrast. This plumage requires seven years for complete transformation. Prior to this the male has a greenish plumage, like the female: during their first three years, they cannot be distinguished. But the fourth year brings subtle changes to the plumage — the bill starts to turn a pale yellowish green and the blue-black feathers begin to appear.

Like the Golden Bowerbird, Satin Bowerbird males gather into clans of up to 20 individuals. Each male defends his territory all year though he shows no qualms about raiding the territories of his neighbours. The bower consists of two parallel walls of sticks, which form the avenue — their bases attached to a mat of woven sticks which lies flat on the ground. On the mat, around the avenue walls, the male places a number of blue objects, remarkable in the similarity of their shades. Before European man, Satin Bowerbirds used natural objects such as flowers and feathers, particularly the tail feathers of the Crimson Rosella. These are still used, but generously supplemented with such man-made objects as bottlecaps, ribbons and — a notable favourite — plastic drinking straws. So strong is the attraction to this shade of blue that Satin Bowerbirds are known to kill and pluck appropriately coloured caged birds. Those who have experimented by removing all the blue objects from the bower and substituted items of other colours have had no success. The bower owner always removes them — and with great haste.

In addition to the blue objects placed around the base of the sides of the avenue, the walls themselves are 'painted'. Using charcoal or a plant material mixed with saliva, the male bowerbird uses his bill as a brush to dab the mixture on the upright sticks. Blackboard chalk has been used as a modern substitute.

Raiding males tear down the bowers of rivals and/or steal their blue objects. There is some evidence that the quantity of blue decorations affects the success of the male in attracting mates. Bower owners take time for maintenance during the breeding season or may tear their own bower down themselves, move the sticks elsewhere and build again. When a female approaches, attracted by his advertising calls, the male goes into his display. He does a stiff-legged hop around the bower, holding his body rigid and pointing his bill downwards. At intervals he flicks his wings outwards. He picks blue objects from the bower and carries them in his bill during the display. At the height of the proceedings, his eyes take on a vivid lilac-purple flush and bulge outwards. Throughout this ritual, the actions are accompanied by whirring, chattering calls. If the female is receptive, they mate quickly in or near the avenue. Once completed, the male aggressively drives her out of the territory. The brief interlude of courtship and mating is the entire duration that the sexes are together. Afterwards he returns to the tasks of repelling intruders, repairing the bower and courting more females. An adult male will attempt as many matings as possible. Meanwhile the female constructs a saucer-shaped nest of leaves and twigs. She lays two to three eggs which require about three weeks to hatch. Many people mistake the bower for the nest of these birds. It should be remembered that each fulfills a completely separate function in the reproductive biology of this species.

Non-breeding males use the long period before maturity to learn about bowers. Black males will tolerate their presence in the territory, so the younger birds watch construction activities, and then make their own practice attempts at bower building. Bowers showing varying stages of completion can be found in the vicinity of the principal bower. Winter flocks — mostly of green birds — gather after the breeding season. They can comprise more than 100 birds, which roam in search of food, dispersing from rainforest and wet eucalypt forest into open woodland. Fruit is the major food item and this leads to raids on orchards and other cultivated crops. In parts of its range, the Satin Bowerbird is now extinct due to the pressure of human settlement and to persecution for its attacks on orchards. Leaves and small animals are also eaten. Satin Bowerbirds will kill and eat small birds without hesitation if they are able to catch them.

These bowerbirds have an array of chatters, hisses, harsh growls and a strange, but highly characteristic metallic buzz, all mixed with mimicry of other species.

Two widely separated populations of Satin Bowerbirds occur along the east coast and adjacent ranges. Individuals of the northern population, in the highlands above 500 metres, are smaller than those from the southeast.

Distribution Satin Bowerbird.

Distribution Golden Bowerbird (darker, northern area) and Regent Bowerbird.

Regent Bowerbird

Though the Regent Bowerbird also builds an avenue, it has not been as well-studied. Unlike the Satin Bowerbird, which places its bower in small clearings on the forest floor, this species conceals its bower amid dense patches of vegetation. Few have been found and observed during courtship. Regent Bowerbirds do not seem to take as much care as the Satin Bowerbirds when building bowers and the results are more unkempt. Snail shells, leaves and coloured berries are used to decorate the bower. Males paint the walls with a yellow mixture made from crushed leaves or fruit pulp. Like their relative, bower owners make raids on each other. Satin Bowerbirds are known to visit the bowers of the Regent Bowerbird, sometimes straightening the walls or adding blue decorations.

The male Regent Bowerbird's courtship dance for a female is much like that of the Satin Bowerbird's — stiff-legged hops, wing spreading and a constant chattering. And as with most bowerbirds, the female assumes all responsibility of rearing the young. Young birds of both sexes resemble adult females but they have dark eyes instead of yellow ones. This change in the colour of their eyes is a sign of approaching maturity which, in the case of males, is accompanied by the gradual development of a

An adult male Satin Bowerbird *Ptilonorhynchus violaceus*.

Robert Eddion 1962.

Bower of the Regent Bowerbird.

yellow bill. Immature males need four to five years to acquire the unmistakable appearance of the black and gold adult.

All the bowerbirds have attractive eggs, one of the most beautiful of which belong to this species. Dark scribbles and lines intertwine across a creamy background, underlaid with light splotches of brown and lavender.

Regent Bowerbirds are shy by nature and this has contributed to the difficulty of studying them at the bower. In certain areas, however, they can become very tame if fed regularly and may even enter an open window in search of handouts.

Green and Spotted Catbirds

The Green Catbird *Ailuroedus crassirostris* takes its name from its colour and from its eerie cat-like call. This sound, which has also been described as resembling a baby crying, is not the only call this bird makes but it is certainly the most diagnostic and easily recognised. Heard most frequently at dawn and dusk, and intermittently during the day, the call is produced by both male and female to advertise their territory to neighbouring catbirds. This species

is more vocal when nesting than during the winter.

Breeding starts about September when the male and female initiate their courtship. Their ritual consists of the slightly larger male chasing the female through the territory. As they go, the chase is interrupted by stops when each bird gives its cat wail and bobs the head and body.

Catbirds are bowerbirds but have major differences from other species in their breeding biology. Most bowerbirds are promiscuous, one male briefly mating with as many females as possible. He defends his territory but gives no assistance in nest building, incubation or rearing of the young. Catbirds are monogamous; they form pairs which remain together through the breeding season. They do not build a bower, but they apparently use a stage which is comprised of bare patches of ground on which broad leaves are placed upside down. Our knowledge of this behaviour in the Green Catbird comes primarily from birds in aviaries, it having been little observed in the wild. The male has a display perch near the stage from which he gives one or more clicking noises followed by three louder cries.

The evolutionary significance of these variations from normal bowerbird behaviour has been

Regent Bowerbird nest.

161

debated. Some scientists consider the catbirds to be the most primitive of the bowerbirds because they have not yet developed bower building or the associated forms of social behaviour. Other ornithologists believe catbirds to be the most advanced members of the family. They assume a secondary loss of bower behaviour and a reversion to monogamy.

The nest is a bulky open cup of leaves, vines and small branches, lined with leaves and moss. It may be anywhere from 5 metres to 30 metres from the ground, wedged in the fork of a tree, or in the crown of a tree fern, or in a thick tangle of vines. Two plain creamy-white eggs are laid. The male defends the nest and a 10 metre radius around it. Unlike most bowerbirds he assists with the care of the chicks. If the nest is approached by a predator, the parents attempt to distract by feigning injury. They flop about on the ground as if hurt, slowly luring the intruder to a safe distance from the nest.

Catbirds eat large amounts of vegetable matter, particularly fruit and leaves, but they will readily eat small animals, with a distinct predilection for nestling birds. Their fondness for fruit leads them to make occasional raids on orchards. Mostly, however, they are birds of the rainforest and rarely leave it. Being sedentary, they remain in the same general area through the year — one banded individual was recaptured at the same spot nine years later! It is not known if a pair of catbirds remain together for more than one season but after breeding ends in January, these birds form flocks of three to ten — occasionally as many as 20 — which roam through the forest in search of food. Pair formation begins the following season.

In the rainforests of the Atherton Tableland district and northern Cape York Peninsula, another species, the Spotted Catbird *A. melanotis,* replaces its southern relative. It also occurs in parts of New Guinea. Slightly smaller than the Green Catbird but more boldly marked, this bird has much the same behaviour but is perhaps even more confined to the rainforest. The call is described as less cat-like than that of its relative. A third species, the White-eared Catbird *A. buccoides,* is limited to New Guinea.

Distribution
Spotted Catbird
(darker area only)
and Green Catbird.

Tooth-billed Catbird

The infrequent and rudimentary stage of the Green Catbird is taken to much more proficient production by the Tooth-billed Catbird *Scenopoeetes dentirostris,* also called Stagemaker because of this behaviour. A male clears an area one to three metres wide on the rainforest floor, removing leaf litter and other debris. On the prepared stage, he spreads large green leaves, always upside down. So particular is this last requirement that he continually fusses to maintain the leaves in this position. If an observer turns the leaves over, the stage owner will quickly return them to their place. Near the tips of its upper and lower bills are teeth-like double notches which earn this species its name. These 'teeth' are used to snip fresh leaves for stage decoration and for food.

Extending across the stage, anywhere from 5 metres to 6 metres above the ground, is a horizontal branch. A courting male spends most of his day singing from this perch, body erect and bill pointed upwards. The song is a medley of different sounds

Green Catbird at nest.

Opposite: Green Catbird *Ailuroedus crassirostris.*

Play arena of the Tooth-billed Catbird.

including harsh scratchings, melodious whistles and low chuckling noises — one noise has been likened to the sound of cloth being torn. During the breeding season, these loud calls make it easy for an observer to find a singing male. Each stage is constructed and guarded by a single male but several individuals may form a loose clan. Where this bird is abundant, stages may be found every 100 metres. Males start and stop while singing, pausing every few seconds to listen to their neighbours' responses. In addition to their courtship song, Tooth-billed Catbirds are said to imitate the songs or other birds, though this has been disputed by some authors. Like the Green Catbird, this species forms monogamous pairs who share breeding duties.

The nest of this species has the reputation of being notoriously hard to find. It is hidden in a thick mass of vines and vegetation, 3 to 25 metres above the ground. The female lays two creamy or buff eggs devoid of markings.

During the non-breeding season from January to July the loud song of the Tooth-billed Catbirds is rarely heard. Tooth-billed Catbirds sometimes join other bowerbirds at fruiting trees but most of the time remain solitary. The species has an unusual distribution: It is found only in the rainforest of the Atherton Tableland district, like several other species but it is further restricted to certain altitudes. Generally Tooth-billed Catbirds can be found only between 600 metres to 1,400 metres although descending to 350 metres in some very wet areas. It may even, though rarely, wonder to the coast.

The nearest relatives to the Tooth-billed Catbird are the other catbirds, the major differences being the much advanced stage construction of the former. These birds are perhaps more closely related than once thought; the occasional attempts at stage building by the Green Catbird have only been known in the last few years.

Distribution Tooth-billed Catbird.

Opposite: Tooth-billed Catbird *Scenopoeetes dentirostris*.

Robert Edden 1983

Robert Folden 1983

Chapter 37

BLACK BUTCHERBIRD

(Cracticidae; Passeriformes)

Butcherbirds have very contrasting natures. On one hand they are rated as having among the best voices of any Australian bird. Their clear, flute-like warbles have deservedly earned them this reputation. However, this pleasing and attractive aspect is quite at odds with its fierce predatory disposition towards small animals, including other birds. Much of their prey is captured on the ground where it is eaten or carried back to a perch. If the victim is too large to eat, the butcherbird wedges it into a forked stick to secure it while it tears it into manageable pieces with the strongly hooked beak. Cage birds are an unfortunate item in the diet of these birds. Sticking its bill between the bars of the cage, the butcherbird grabs its victim and then pulls it out; if the smaller bird does not initially fit through the bars the butcherbird reduces it to a more suitable size. Handouts around picnic grounds and houses are taken by individuals used to humans. The occasional seeds round out the diet.

Most butcherbirds are black and white but one, the Black Butcherbird *Cracticus quoyi* is, as the name indicates, all black. This species is found on both sides of the Torres Strait. Of the four Australian Butcherbirds, it is the only one that occupies rainforest, though it is also found in mangroves, monsoon forest and in more open situations not far from these habitats. The other species replace it in the drier and less densely vegetated parts of the country. Even within the thick forest, the Black Butcherbird spends most of its time near the edges. Here it skulks among the leaves. It is generally shy and wary though some birds become tame round houses or parks, boldly approaching people to solicit leftover food scraps.

The beautiful song 'cou cou cooka cooka' is given frequently, but most often during the height of breeding , when both parents engage in duets. Black Butcherbirds are sedentary and pairs maintain territories throughout the year. Starting in September, the adults construct the nest, a large bowl of sticks and twigs placed in the fork of a tree five to ten metres above the ground. The female lays two to three eggs; these are dull reddish pink or grey-green with darker spots and blotches. Both adults care for the young and are often aided by offspring from earlier broods which remain in the territory for some time after becoming independent. Butcherbirds fiercely defend the nest against intruders, swooping on them and attacking with their formidable bills.

Distribution Black Butcherbird.

Black Butcherbirds are usually observed in pairs or small family parties. Males are larger than females but they are otherwise identical. In the middle of the eastern Queensland population, some birds differ by being a dark rufous brown rather than black. These are immatures. Though other individuals of the same age are black, both colour phases may occur in a single brood. Breeding by rufous individuals has been recorded.

Opposite: Black Butcherbird *Cracticus quoyi*.
